Kyle
xu

Fun Math Problem Solving
For Elementary School

Solutions Manual

David Reynoso
John Lensmire
Kevin Wang
Kelly Ren

Kyle
xy

TITLES PUBLISHED BY ARETEEM INSTITUTE

Cracking the High School Math Competitions (and Solutions Manual) - Covering AMC 10 & 12, ARML, and ZIML

Mathematical Wisdom in Everyday Life (and Solutions Manual) - From Common Core to Math Competitions

Geometry Problem Solving for Middle School (and Solutions Manual) - From Common Core to Math Competitions

Fun Math Problem Solving For Elementary School (and Solutions Manual)

ZIML Math Competition Book Division M 2016-2017

ZIML Math Competition Book Jr Varsity 2016-2017

COMING SOON

ZIML Math Competition Book Division E 2016-2017

ZIML Math Competition Book Division H 2016-2017

ZIML Math Competition Book Varsity 2016-2017

Counting & Probability for Middle School (and Solutions Manual) - From Common Core to Math Competitions

Number Theory Problem Solving for Middle School (and Solutions Manual) - From Common Core to Math Competitions

The books are available in paperback and Kindle eBook formats. To order the books, visit https://areteem.org/bookstore.

ISBN: 1-944863-08-7
ISBN-13: 978-1-944863-08-1

First printing, October 2017.

Contents

Introduction

This book is part of the ongoing effort by Areteem Institute to inspire students, parents, and teachers to gain a deeper understanding and appreciation of mathematics. This book is aimed for students in 3rd, 4th, and 5th grade in elementary school. This book leads readers through complex math concepts via age-appropriate approaches, such as fun stories in real-life scenarios, riddles and puzzles, magic tricks, cartoon drawings, jokes, etc. Math is fun! The authors of the book are experts in math who are passionate educators and they work hard to present the fun aspect of math to young students to stimulate interest in math and develop problem solving and critical thinking skills at an early age. In addition, this book reviews and expands state math standards, including the Common Core Standards, particularly the Operations and Algebraic Thinking (OA), Numbers and Operations in Base Ten (NBT), and Measurement and Data (MD) domains at the 3rd, 4th, and 5th grade level.

This book is the Solutions Manual of the accompanying Student Workbook, *Fun Math Problem Solving For Elementary School*. The Student Workbook contains all the material and practice problems, and answers to all practice problems. The Solutions Manual includes in-depth solutions to all of the quick response and practice problems.

The problems in this book offer the student a chance to start developing problem solving techniques that will be useful not only in mathematics but also in everyday life.

While most of the problems in the last few chapters of this book could be solved using algebraic techniques, we strongly encourage the students to attempt solving them without algebra. Not only the solutions are easier to understand, but students gain a deeper understanding of the problems this way, and the students develop logical thinking skills through creative reasoning. Some students may think that in solving these problems "we are doing the same as if we were solving equations", however, when solving equations we usually do not identify what the middle steps of the solving process mean for the problem itself, which may play a huge part into understanding basic abstract mathematical concepts.

About Areteem Institute

Areteem Institute is an educational institution that develops and provides in-depth and advanced math and science programs for K-12 (Elementary School, Middle School, and High School) students and teachers. Areteem programs are accredited supplementary programs by the Western Association of Schools and Colleges (WASC). Students may attend the Areteem Institute through these options:

- Live and real-time face-to-face online classes with audio, video, interactive online whiteboard, and text chatting capabilities;
- Self-paced classes by watching the recordings of the live classes;
- Short video courses for trending math, science, technology, engineering, English, and social studies topics;
- Summer Intensive Camps on prestigious university campuses and Winter Boot Camps;
- Practice with selected daily problems for free, monthly ZIML competition at `http://ziml.areteem.org`.

The Areteem courses are designed and developed by educational experts and industry professionals to bring real world applications into STEM education. The programs are ideal for students who wish to build their mathematical strength in order to excel academically and eventually win in Math Competitions (AMC, AIME, USAMO, IMO, ARML, MathCounts, Math Olympiad, ZIML, and other math leagues and tournaments, etc.), Science Fairs (County Science Fairs, State Science Fairs, national programs like Intel Science and Engineering Fair, etc.) and Science Olympiad, or purely want to enrich their academic lives by taking more challenges and developing outstanding analytical, logical thinking and creative problem solving skills.

Since 2004 Areteem Institute has been teaching with methodology that is highly promoted by the new Common Core State Standards: stressing the conceptual level understanding of the math concepts, problem solving techniques, and solving problems with real world applications. With the guidance from experienced and passionate professors, students are motivated to explore concepts deeper by identifying an interesting problem, researching it, analyzing it, and using a critical thinking approach to come up with multiple solutions.

Thousands of math students who have been trained at Areteem achieved top honors and earned top awards in major national and international math competitions, including Gold Medalists in the International Math Olympiad (IMO), top winners and qualifiers at the USA Math Olympiad (USAMO/JMO), and AIME, top winners at the Zoom

International Math League (ZIML), and top winners at the MathCounts National. Many Areteem Alumni have graduated from high school and gone on to enter their dream colleges such as MIT, Cal Tech, Harvard, Stanford, Yale, Princeton, U Penn, Harvey Mudd College, UC Berkeley, UCLA, etc. Those who have graduated from colleges are now playing important roles in their fields of endeavor.

Further information about Areteem Institute, as well as updates and errata of this book, can be found online at http://www.areteem.org.

Acknowledgments

This book contains many years of collaborative work by the staff of Areteem Institute. This book could not have existed without their efforts. The materials in this book were prepared by Kelly Ren and Kevin Wang for Areteem's Young Math Olympians courses, and were later updated and expanded by John Lensmire and David Reynoso. Especially, the illustrations in each chapter were created by David. Yes, the same David Reynoso who is a mathematician by trade and artist by nature!

The examples and problems in this book were either created by the Areteem staff or adapted from various sources, including other books and online resources. We extend our gratitude to the original authors of all these resources.

Last but not least, special thanks go to Saber and Hazel, who starred in the pictures on the covers and chapter images, photographed by Kelly Ren.

1. It Matters Where It Is

Problem 1.1 **What numbers are shown below?**

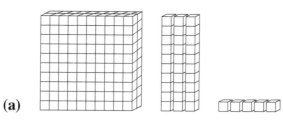

(a)

Answer

235

Solution

$2 \times 100 + 3 \times 10 + 5 \times 1 = 235$

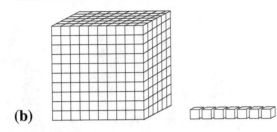

(b)

Answer

607

Solution

$6 \times 100 + 7 \times 1 = 607$

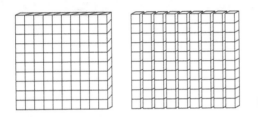

(c)

Answer

192

Solution

$1 \times 100 + 9 \times 10 + 2 \times 1 = 192$

Problem 1.2 **Do you get an odd number or even number if you add 10 odd numbers?**

Answer

Even

Solution

When we add 2 odd numbers we get an even number. This time we are adding 10 odd numbers, that is, 5 pairs of odd numbers. So, the sum will always be even.

Problem 1.3 **Can you find three numbers whose sum is 100 with only one of the**

numbers odd? Why or why not?

No

Solution

The sum of an odd number and two even numbers will always be odd, but 100 is even. So it is impossible to find three numbers that add up to 100 only one of them being odd.

Problem 1.4 **Do you get an odd or even number if you multiply an odd number by an even number?**

Answer

Even

Solution

Multiplying by an even number always gives us an even number. It does not matter if the other number is even or odd.

Problem 1.5 **Add any two odd numbers. The ones digit of the sum is always _____.**

Answer

Even

Solution

We know the sum of two odd numbers is always an even number. Even numbers end with even digits, so the ones digit of the sum of two odd numbers will always be even.

Problem 1.6 **In 98960, which digit is the tens place?**

Answer

6

Solution

The tens digit is the one in the second to last place. In this case that is 6.

Problem 1.7 What is the sum of the digits in the number one million?

Answer

1

Solution

When we write one million, we have a digit 1 followed by six 0s. So, the sum of their digits will be $1+0+0+0+0+0+0 = 1$.

Problem 1.8 My aunt owns many cars. If the number of cars my aunt owns is 1 less than the thousands digit of 17854, how many cars does she own?

Answer

6

Solution

The thousands digit of 17854 is 7, and one less than 7 is 6.

Problem 1.9 Fill in the blanks

(a) $5000 = $ _____ **hundreds** $= $ _____ **ones**

Answer

50; 5000

Solution

$5000 = 50 \times 100$, so it is equal to 50 hundreds; and $5000 = 5000 \times 1$, so it is also equal to 5000 ones.

(b) $3000 = $ _____ **hundreds** $= $ _____ **ones**

Answer

30; 3000

Solution

$3000 = 30 \times 100$, so it is equal to 30 hundreds; and $3000 = 3000 \times 1$, so it is also equal to 3000 ones.

(c) $3300 =$ _____ **hundreds** $=$ _____ **ones**

Answer

33; 3300

Solution

$3300 = 33 \times 100$, so it is equal to 33 hundreds; and $3300 = 3300 \times 1$, so it is also equal to 3300 ones.

Problem 1.10 What number has

(a) 12 **hundreds and** 9 **ones?**

Answer

1209

Solution

A number that has 12 hundreds and 9 ones is the same as $12 \times 100 + 9 \times 1 = 1209$.

(b) 2 **thousands,** 21 **tens, and** 4 **ones?**

Answer

2214

Solution

A number that has 2 thousands, 21 tens and 4 ones is the same as $2 \times 1000 + 21 \times 10 + 4 \times 1 = 2214$.

(c) 65 **ones and** 3 **thousands?**

Answer

3065

Solution

A number with 3 thousands and 65 ones is the same as $3 \times 1000 + 65 \times 1 = 3065$.

Problem 1.11 Fill in the blanks.

(a) _____ $+ 300 +$ _____ $+ 7 = 2397$

Answer

2000 and 90

Solution

2397 is the same as $2000 + 300 + 90 + 7$, so we are missing 2000 and 90 in the sum.

(b) $2 +$ _____ $+ 800 + 9000 = 9832$

Answer

3

Solution

9832 is the same as $9000 + 800 + 30 + 2$, so we are missing a 30 in the sum.

(c) _____ $+ 700 +$ _____ $+ 20 = 9728$

Answer

9000 and 8

Solution

$9728 = 9000 + 700 + 20 + 8$, so we should have 9000 and 8 in the blank spaces.

(d) $8 + 80 +$ _____ $+ 8000 = 8888$

Answer

800

Solution

8888 has 8 thousands, 8 hundreds, 8 tens and 8 ones. We are missing 8 hundreds, so 800.

(e) _____ $+ 80 + 800 = 884$

Answer

4

Solution

884 is the same as $800 + 80 + 4$, so we need a 4 in the blank space.

Problem 1.12 Fill in the blanks.

(a) $8 + 400 + 8000 + 30 =$ _____

Answer

8438

Solution

The number will have 8 ones, 4 hundreds, 8 thousands, and 3 tens. So, the number is 8438.

(b) $3 + 0 + 7000 + 10 =$ _____

Answer

7013

Solution

The number will have 3 ones, 7 thousands, 1 ten, and no hundreds. The number is 7013.

(c) $8 + 60 + 600 + 7000 = $ _____

Answer

7668

Solution

The number will have 8 ones, 6 tens, 6 hundreds and 7 thousands. The number is 7668.

Problem 1.13 **Find the number that has**

(a) **0 thousands, 3 hundreds, 3 tens, and 6 ones**

Answer

336

Solution

The number we are looking for will be the same as $0 \times 1000 + 3 \times 100 + 3 \times 10 + 6 \times 1 = 336$.

(b) **2 thousands, 0 hundreds, 0 tens, and 6 ones**

Answer

2006

Solution

The number we want is $2 \times 1000 + 0 \times 100 + 0 \times 10 + 6 \times 1 = 2006$.

(c) **5 thousands, 0 hundreds, 5 tens, and 3 ones**

Answer

5053

Solution

The number is the same as $5 \times 1000 + 0 \times 100 + 5 \times 10 + 3 \times 1 = 5053$

Problem 1.14 What 3-digit numbers number have 1 ten and 1 more one than tens?

Answer

112, 212, 312, 412, 512, 612, 712, 812 and 912

Solution

A number that has 1 ten and 1 more one than tens will always end in 12. We want 3-digit numbers, so we just need to put in the first digit. The first digit can be any digit other than 0, so the possible numbers we have are 112, 212, 312, 412, 512, 612, 712, 812 and 912.

Problem 1.15 What number has 0 thousands, 8 hundreds, 1 more ten than hundreds, and no ones?

Answer

890

Solution

Since the number has 1 more ten than hundreds, and it has 8 hundreds, it has 9 tens. So the number we want is $800 + 90 = 890$.

Problem 1.16 If the ones digit of a number is 4, the tens digit is twice the ones digit, and the hundreds digit is half of the ones digit, what is the number?

Answer

284

Solution

The tens digit of our number will be $2 \times 4 = 8$, and the thousands digit will be $4 \div 2 = 2$. So the number is 284.

Problem 1.17 What is the difference between the largest and the smallest 5-digit numbers, each greater than 40000, that can be formed using only the digits 2, 3, 4, 5 and 6 if no digit is used more than once?

Answer

23076

Solution

The largest number using the given digits is 65432. If the smaller number needs to be greater than 40000, it must start with a 4. Hence the smallest is 42356. This gives a difference of $65432 - 42356 = 23076$.

Problem 1.18 What 4-digit number has 2 tens, 6 hundreds, 1 more than the hundreds digit on the thousands digit, and 2 less ones than hundreds?

Answer

7624

Solution

The thousands digit of the number will be 1 more than 6, so it will be 7. The ones digit will be 2 less than 6, so it will be 4. This means the number is 7624.

Problem 1.19 Find a 3-digit number such that the tens digit is a multiple of the hundreds digit and the ones digit is a multiple of the tens digit.

Answer

Sample answer: 248

Solution

There are multiple correct answers to this problem. To verify if a number works, we just need to make sure that we can divide the last digit by the second digit and that we can divide the second digit by the first digit. Some other sample answers are: 100, 111, 224, 366, 444. etc.

Problem 1.20 A number has 4 digits. Its ones digit is the second highest one digit number. The tens digit is half the ones digit. The hundreds digit is one more than half of the tens digit. The thousands digit is the difference between the tens digit and the hundreds digit. What is the number?

Answer

1348

Solution

The second highest one digit number is 8, so that will be the ones digit. Half of 8 is $8 \div 2 = 4$, so that will be the tens digit. One more than half of 4 is $2 \div 2 + 1 = 3$, so that will be the hundreds digit. The difference between the tens digit and the hundreds digit is $4 - 3 = 1$, so that will be the thousands digit. The number is 1348.

Problem 1.21 There is a 3-digit number. Its ones digit is 5 times its hundreds digit. Its tens digit is the sum of its ones digit and its hundreds digit. What is the 3-digit number?

Answer

165

Solution

If the hundreds digit is 1, then the ones digit is $5 \times 1 = 5$. If the hundreds digit is bigger than 1, then 5 times that digit will be bigger than 9, which is impossible as a digit. This means the hundreds digit is 1 and the hundreds digit is 5. The tens digit will then be $1 + 5 = 6$ so the number is 165.

Problem 1.22 There is a 2-digit number. The sum of the two digits is 7. If you switch the two digits, the new number is 27 more than the original 2-digit number. What is the original 2-digit number?

Answer

25

Solution

The two digit numbers that whose digits add up to 7 are 16, 25, 34, 43, 52, 61 and 70. By switching the digit of each of those numbers we can see that 52 is 27 more than 25. The original number is 25.

Problem 1.23 What is the difference between the greatest 5-digit number and

the smallest 3-digit number?

Answer

99899

Solution

The greatest 5-digit number is 99999. The smallest 3-digit number is 100. The difference between those two numbers is $99999 - 100 = 99899$.

Problem 1.24 **Write the greatest 4-digit number using all different digits with a 6 in the tens place.**

Answer

9867

Solution

We want to use the biggest digits we can starting from the left, making sure we use 6 in the tens place. We will have 9 thousands, 8 hundreds, and 7 ones. The number is 9867.

Problem 1.25 **We want to write some 7 digit numbers. Let's pretend we can only use the digits 5, 1, 6, 2, 0, 9, and 7, and we can only use each digit once.**

(a) What is the largest number we can write?

Answer

9765210

Solution

We want to make sure to use the biggest digits first. If we order the digits we have available from biggest to smallest, they are 9, 7, 6, 5, 2, 1 and 0. The number is 9765210.

(b) The smallest number we can write?

Answer

1025679

Solution

We want to make sure to use the smallest digits first, but we need to make sure that we use something other than 0 for the first digit. The digits in order from smallest to largest are 0, 1, 2, 5, 6, 7, 9. We will have to use 1 first so the number has seven digits. So the number is 1025679.

(c) The largest odd number?

Answer

9765201

Solution

We just need to make sure to have the last digit to be odd and make sure to use the smallest digit we can for that. 1 is the smallest odd digit we have available, so the number is 9765201.

(d) The smallest even number?

Answer

1025796

Solution

We want to make sure to have the last digit to be an even number and we want to use the biggest digit we can. In the number we found to be the smallest, we can move the 6 to the last place (since that is the biggest even digit we have available) so the number will be 1025796.

Problem 1.26 Use the digits 0, 2, 3, 5, 7 to make two 5-digit numbers such that the difference of the two numbers is as large as possible. What is the difference?

Answer

54963

Solution

To have the difference be as large as we can we want to make sure to subtract the smallest number we can make from the biggest number we can make. The largest 5-digit number we can make with the digits we have available is 75320, and the smallest 5-digit number we can make is 20357. Their difference is $75320 - 20357 = 54963$.

Problem 1.27 Uncle Jim got lost while we were driving back to California from Montréal. He saw a sign that said how many miles we were from Los Angeles. Since he was driving fast, he couldn't quite see the number, but he knew it had 4 digits. I saw the exact number, but I wanted to have some fun so instead of telling him the number right away I gave him some clues:

- The number has the digit 1 somewhere.
- The digit in the hundreds place is three times the digit in the thousands place.
- The digit in the ones place is 4 times the digit in the tens place.
- The thousands digit is 2.

How far away are we from Los Angeles?

Answer

2614 miles

Solution

The last clue tells us that the thousands digit is 2. The second clue tells us that the hundreds digit is $3 \times 2 = 6$. Clue number three tells us that the ones digit must be 4 times the thousands digit, and we haven't used 1 as the first clue is telling us to, so the thousands digit must be 1 and the ones digit is $1 \times 4 = 4$. The distance to Los Angeles is 2614 miles.

Problem 1.28 George's family lives in a house with a 4-digit street number. The difference between the first digit and the last digit is 8. The 2nd digit is twice the first digit, and the 3rd digit is twice the 2nd digit. What is the street number of George's house?

Answer

1249

Solution

The first and last digits must be 1 and 9, since the is the only way we can have that their difference be 8. If 9 was the first digit, we would not be able to find the second digit, because $9 \times 2 = 18$ is no longer a single digit number. So, the first digit is 1 and the last digit is 9. This way, the second digit is $1 \times 2 = 2$ and the third digit is $2 \times 4 = 4$. Thus the number of George's house is 1249.

Problem 1.29 There were as many kids in the first car of the roller coaster as the largest possible sum of two different 1-digit numbers. How many kids were in the first car of the roller coaster?

Answer

17

Solution

If we want the largest sum we should use the largest digits we can. We want them to be different so we need to use 9 and 8. Their sum is $9 + 8 = 17$.

Problem 1.30 The number of pieces of candy I ate last year is the same as the largest even number smaller than 1500. How many pieces of candy did I eat?

Answer

1498

Solution

1500 is an even number, so the largest even number smaller than it has to be 2 less than 1500. That means I ate $1500 - 2 = 1498$ pieces of candy.

Problem 1.31 I have a toy car collection that has as many cars as the smallest odd number bigger than 3900. How many cars do I have in my collection?

Answer

3901

Solution

Since 3900 is an even number, the smallest odd number bigger than it is just 1 more. So there are $3900 + 1 = 3901$ cars in my collection.

Problem 1.32 Ron wants to go to the arcade so he will need as many quarters as possible. His dad gave him two $5 bills, and his mom gave him three $1 bills and one $10 bill. How many quarters can he get with all the money he has?

Answer

92

Solution

Since we know how many of each type of bill Ron has, we know that he has $2 \times 5 + 3 \times 1 + 1 \times 10 = 23$ dollars in total. For each dollar, Ron can get 4 quarters, so he can get $23 \times 4 = 92$ quarters.

Problem 1.33 Travis has been saving all the change he gets whenever he uses a $1 bill. After one month he has gathered 150 pennies, 53 nickels, and 7 quarters. Does he have enough money to buy a popsicle that costs $5.99?

Answer

No

Solution

We can figure out how much money Travis has in total by using the face value of each of the coins he has. A penny is worth 1¢, so $0.01. A nickel is worth 5¢, so $0.05. A quarter is worth 25¢, so $0.25. This means Travis has $150 \times 0.01 + 53 \times 0.05 + 7 \times 0.25 = 5.90$ dollars. He has 9¢ less than what he needs to buy the popsicle, so he does not have enough money.

Problem 1.34 Can you find 5 consecutive 2-digit numbers such that their sum has 6 tens and no hundreds?

Answer

10, 11, 12, 13, and 14, or 11, 12, 13, 14, and 15

Solution

If we add the first 5 consecutive 2-digit numbers we have that their sum is $10 + 11 + 12 + 13 + 14 = 60$, which is a number with 6 tens and no hundreds. If instead of 10 we start with 11, we get $11 + 12 + 13 + 14 + 15 = 65$ which is also a number with 6 tens and no hundreds. If we do from 12 to 16 we get $12 + 13 + 14 + 15 + 16 = 70$ which is no longer a number with 6 tens. This means the only two possible solutions are 10, 11, 12, 13, and 14, or 11, 12, 13, 14, and 15.

Problem 1.35 Find three consecutive numbers with 2 hundreds whose sum is even.

Answer

Sample answer: 201, 202 and 203

Solution

Any three consecutive numbers with the middle number being even will work. We just need to make sure each of them has 2 hundreds.

2. Mathemagics

Problem 2.1 Write 5 pairs of 1-digit numbers such that the sum of each pair is 10.

Answer

1 & 9; 2 & 8; 3 & 7; 4 & 6; 5 & 5; ...

Problem 2.2 Write 5 pairs of 1-digit numbers such that the sum of each pair is 9.

Answer

1 & 8; 2 & 7; 3 & 6; 4 & 5; 5 & 4; ...

Problem 2.3 Use the pairs you found in the previous question to write 10 pairs of 2-digit numbers that add up to 99. Try to be creative and include numbers that start with different digits!

Answer

Sample answers: 64 & 35; 12 & 87

Problem 2.4 Write 10 pairs of 2-digit numbers, that don't end in 0, such that the sum of each pair is 100.

Answer

Sample answers: 67 & 33; 29 & 71

Problem 2.5 Write 10 pairs of 3-digit numbers, that don't end in 00, such that the sum of each pair is 1000.

Answer

Sample answers: 564 & 436; 783 & 217

Problem 2.6 Group the numbers 6, 9, 11, and 8 into pairs so that each pair has the same sum.

Answer

6 & 11; 8 & 9

Solution

$6 + 11 = 17$; $8 + 9 = 17$

Problem 2.7 Find the product of the following "teens".

(a) 19×12

Answer

228

Solution

$9 \times 2 = 18$ and $10 + 1 + 9 + 2 = 22$ $\Rightarrow$ $19 \times 12 = 228$

(b) 16×14

Answer

224

Solution

$6 \times 4 = 24$ and $10 + 2 + 6 + 4 = 22$ $\Rightarrow$ $16 \times 14 = 224$

(c) 13×18

Answer

234

Solution

$3 \times 8 = 24$ and $10 + 2 + 3 + 8 = 23$ $\Rightarrow$ $13 \times 18 = 234$

Problem 2.8 Multiply the following numbers by 11

(a) 43

Answer

473

Solution

4   4+3   3 $\Rightarrow$ $43 \times 11 = 473$

(b) 63

Answer

693

Solution

6   6+3   3 $\Rightarrow$ $63 \times 11 = 693$

(c) 33

Answer

363

Solution

$3 \quad 3+3 \quad 3 \quad \Rightarrow \quad 33 \times 11 = 363$

Problem 2.9 Find the squares of the following numbers.

(a) 18^2

Answer

324

Solution

$18 + 2 = 20 \quad 18 - 2 = 16 \quad 2^2 = 4 \quad \Rightarrow \quad 18^2 = 20 \times 16 + 4 = 324$

(b) 19^2

Answer

361

Solution

$19 + 1 = 20 \quad 19 - 1 = 18 \quad 1^2 = 1 \quad \Rightarrow \quad 19^2 = 20 \times 18 + 1 = 361$

(c) 17^2

Answer

289

Solution

$17 + 3 = 20 \quad 17 - 3 = 14 \quad 3^2 = 9 \quad \Rightarrow \quad 17^2 = 20 \times 14 + 9 = 289$

Problem 2.10 Let's square some numbers! Try to do it as fast as you can! (This does not mean that we want you to break your pencil because you are writing too fast, but remember we learned a technique to do these calculations quicker!)

(a) 65^2

Answer

4225

Solution

$6 \times 7 = 42 \quad \Rightarrow \quad 65^2 = 4225$

(b) 95^2

Answer

9025

Solution

$9 \times 10 = 90 \quad \Rightarrow \quad 95^2 = 9025$

(c) 75^2

Answer

5625

Solution

$7 \times 8 = 56 \quad \Rightarrow \quad 75^2 = 5625$

Problem 2.11 **All the following pairs of numbers have something in common. See if you can identify what that is and find their sums!**

(a) $70 + 29$

Answer

99

Solution

In all of these problems the digits add up to 9!

(b) $462 + 537$

Answer

999

(c) $468 + 531$

Answer

999

(d) $7755 + 2244$

Answer

9999

(e) $8252 + 1747$

Answer

9999

Problem 2.12 Add up the following numbers. Try to identify pairs of numbers that add up to $9, 99, 999, 9999, 10, 100, 1000$ **or** 10000 **to make your job easier!**

(a) $326 + 674 + 54 + 45$ **(Find a pair of numbers that adds up to** 1000 **and another pair that adds up to** 99**).**

Answer

1099

Solution

$(326 + 674) + (54 + 45) = 1000 + 99 = 1099$

(b) $5694 + 4306 + 598$ **(Find a pair of numbers that adds up to** 10000**).**

Answer

10598

Solution

$(5694 + 4306) + 598 = 10000 + 598 = 10598$

(c) $6845 + 3154 + 4987 + 5013$ (**Find a pair that adds up to** 10000 **and another pair that adds up to** 9999).

Answer

19999

Solution

$(6845 + 3154) + (4987 + 5013) = 9999 + 10000 = 19999$

Problem 2.13 **Times 11 time! Multiply the following numbers by 11. Remember to add up neighbors to save time!**

(a) 743×11

Answer

8173

Solution

7   7+4   4+3   3 $\Rightarrow$ 7   11   7   3 $\Rightarrow$ 8173

(b) 6172×11

Answer

67892

Solution

6   6+1   1+7   7+2   2 $\Rightarrow$ 6   7   8   9   2 $\Rightarrow$ 67892

(c) 3241×11

Answer

35651

Solution

$3 \quad 3+2 \quad 2+4 \quad 4+1 \quad 1 \quad \Rightarrow \quad 3 \quad 5 \quad 6 \quad 5 \quad 1 \quad \Rightarrow \quad 35651$

Problem 2.14 Some of the numbers in the following questions are close to numbers that are easier to work with. Use that to your advantage when calculating!

(a) $1456 - 299$

Answer

1157

Solution

$1456 - 299 = 1456 - 300 + 1 = 1156 + 1 = 1157$

(b) $19998 + 3 + 1999 + 998 + 3 + 999$

Answer

24000

Solution

$$
\begin{aligned}
& 19998 + 3 + 1999 + 998 + 3 + 999 \\
={} & (20000 - 2) + 3 + (2000 - 1) + (1000 - 2) + 3 + (1000 - 1) \\
={} & 20000 - 2 + 3 + 2000 - 1 + 1000 - 2 + 3 + 1000 - 1 \\
={} & 24000
\end{aligned}
$$

(c) $49996 + 39993 + 29992 + 19991 + 998$

Answer

140970

Solution

$$
\begin{aligned}
& 49996 + 39993 + 29992 + 19991 + 998 \\
=\ & (50000 - 4) + (40000 - 7) + (30000 - 8) + (20000 - 9) + (1000 - 2) \\
=\ & 50000 - 4 + 40000 - 7 + 30000 - 8 + 20000 - 9 + 1000 - 2 \\
=\ & 150000 - 30 \\
=\ & 14970
\end{aligned}
$$

Problem 2.15 Before you add the numbers, find pairs of numbers that add up to something that ends with a 0. That should make the calculation a lot easier!

(a) $191 + 809 + 259 + 2329 + 1741$

Answer

5329

Solution

$(191 + 809) + (259 + 1741) + 2329 = 1000 + 2000 + 2329 = 5329$

(b) $137 + 356 + 863 + 644$

Answer

2000

Solution

$(137 + 863) + (356 + 644) = 1000 + 1000 = 2000$

(c) $829 + 571 - 692 - 308$

Answer

400

Solution

$(829 + 571) - (692 + 308) = 1400 - 1000 = 400$

Problem 2.16 Take a look at the last two digits of the numbers in the following. Find something in them that may help you do this computations quicker.

(a) $673 + 528 - 373 + 472$

Answer

1300

Solution

$(673 - 373) + (528 + 472) = 300 + 1000 + 1300$

(b) $2273 - 655 - 345 - 273$

Answer

1000

Solution

$(2273 - 273) - (655 + 345) = 2000 - 1000 = 1000$

(c) $2948 + 4355 - 648 - 155 + 74$

Answer

6574

Solution

$(2948 - 648) + (4355 - 155) + 74 = 2300 + 4200 + 74 = 6574$

Problem 2.17 Use the strategy for squaring numbers that end up in 5 to find out the squares of this numbers. Remember that the square of a number that ends in 5 always ends in 25! Hint: We also learned a strategy for multiplying "teens" that may help!

(a) 105^2

Answer

11025

Solution

$10 \times 11 = 110 \quad \Rightarrow \quad 105^2 = 11025$

(b) 115^2

Answer

13225

Solution

$11 \times 12 = 132 \quad \Rightarrow \quad 105^2 = 13225$

(c) 125^2

Answer

15625

Solution

$12 \times 13 = 156 \quad \Rightarrow \quad 125^2 = 15625$

(d) 135^2

Answer

18225

Solution

$13 \times 14 = 182 \quad \Rightarrow \quad 135^2 = 18225$

(e) 145^2

Answer

21025

Solution

$14 \times 15 = 210 \quad \Rightarrow \quad 145^2 = 21025$

(f) 155^2

Answer

24025

Solution

$15 \times 16 = 240 \quad \Rightarrow \quad 155^2 = 24025$

(g) 165^2

Answer

27225

Solution

$16 \times 17 = 272 \quad \Rightarrow \quad 165^2 = 27225$

(h) 175^2

Answer

30625

Solution

$17 \times 18 = 306 \quad \Rightarrow \quad 175^2 = 30625$

(i) 185^2

Answer

34225

Solution

$18 \times 19 = 342 \quad \Rightarrow \quad 185^2 = 34225$

Problem 2.18 Fill in the blanks

(a) $78 = 25 \times \underline{\quad} + \underline{\quad}$

Answer

3; 3

Solution

$78 = 25 \times 3 + 3$

(b) $277 + 177 = 25 \times \underline{\hspace{1cm}} + 4 = \underline{\hspace{1cm}}$

Answer

18; 454

Solution

$277 = 25 \times 11 + 2, \quad 177 = 25 \times 7 + 2 \quad \Rightarrow \quad 277 + 177 = 25 \times 18 + 4$

(c) $273 + 152 + 25 = 25 \times \underline{\hspace{1cm}} = 50 \times \underline{\hspace{1cm}} = \underline{\hspace{1cm}}$

Answer

18; 9; 450

Solution

$273 = 25 \times 11 - 2 \quad 152 = 25 \times 6 + 2 \quad 25 = 25 \times 1 \Rightarrow \quad 273 + 152 + 25 = 25 \times 18 = 50 \times 9 = 450$

Problem 2.19 In each of these sums, the summands are each close to a number. Identify that number to fill in the blanks and find the sum.

(a) $12 + 13 + 9 + 8 + 13 + 9 = 10 \times \underline{\hspace{1cm}} + \underline{\hspace{1cm}} = \underline{\hspace{1cm}}$

Answer

6; 4; 64

Solution

$(10+2) + (10+3) + (10-1) + (10-2) + (10+3) + (10-1) = 10 \times 6 + 4 = 64$

(b) $98 + 105 + 94 + 101 + 99 = 100 \times \underline{\hspace{1cm}} - \underline{\hspace{1cm}} = \underline{\hspace{1cm}}$

Answer

5; 3; 497

Solution

$(100-2)+(100+5)+(100-6)+(100+1)+(100-1)=100\times5-3=497$

(c) $997+1004+1003+996+999+1005=1000\times\underline{\quad}+\underline{\quad}=\underline{\quad}$

Answer

6; 4; 6004

Solution

$$(1000-3)+(1000+4)+(1000+3)+(1000-4)+(1000-1)+(1000+5)$$
$$=1000\times6+4=6004$$

Problem 2.20 Identify a number that is close to all the numbers in the sum. Use that number to help find the sum.

(a) $47+56+43+60+43+52+40+48+45+58+41+55$

Answer

588

Solution

$50\times12-3+6-7+10-7+2-10-2-5+8-9+5=600-12=588$

(b) $504+505+510+499+504+492+501+503+490+492+501+499$

Answer

6501

Solution

$500\times13+4+5+10-1+4-8+1+3-10-8+1+1-1=6500+1=6501$

(c) $1505 + 1509 + 1497 + 1499 + 1506 + 1496$

Answer

9012

Solution

$1500 \times 6 + 5 + 9 - 3 - 1 + 6 - 4 = 9000 + 12 = 9012$

Problem 2.21 **In the following sums, pairs of numbers add up to a common number. Use that to your advantage.**

(a) $57 + 13 + 9 + 29 + 41$

Answer

149

Solution

$(57 + 13) + (29 + 41) + 9 = 70 + 70 + 9 = 149$

(b) $70 + 50 + 33 + 12 + 69 + 51 + 108$

Answer

393

Solution

$(70 + 50) + (12 + 108) + (69 + 51) + 33 = 120 + 120 + 120 + 33 = 360 + 33 = 393$

(c) $99 + 13 + 86 + 24 + 16 + 75$

Answer

313

Solution

$99 + 99 + 99 + 16 = 300 + 16 - 3 = 313$

Problem 2.22 Find pairs of numbers that add up to the same number. Then use these numbers to add and subtract quicker. Be careful with the minus signs!

(a) $29 + 46 - 28 - 47 + 38 + 37 + 44 + 31$

Answer

150

Solution

$(29+46) - (28+47) + (38+37) + (44+31) = 75 - 75 + 75 + 75 = 150$

(b) $15 + 47 + 42 + 20 - 30 - 45 - 32 - 17 + 46 + 16$

Answer

62

Solution

$(15+47) + (42+20) - (30+32) - (45+17) + (46+16) = 62 + 62 - 62 - 62 + 62 = 62$

(c) $21 + 17 + 24 + 28 + 23 + 22 - 16 + 18 + 16 + 29 + 27 - 29$

Answer

180

Solution

$$(21+24) + (17+28) + (23+22) - 16 + 16 + (18+27) + 29 - 29$$
$$= 45 + 45 + 45 + 0 + 45 + 0 = 180$$

Problem 2.23 Dulce is going to order an enormous amount of croissants for some upcoming breakfast meetings. Since she is ordering so many croissants, the baker offered her a deal that for every box of croissants she buys she gets 1 extra croissant for free. If croissants normally come in boxes of 10, how many croissant (including the free croissant per box) wlil Dulce get if she orders

(a) 140 **boxes of croissants?**

Answer

1540

Solution

For every box of 10 croissants Dulce buys, she gets 1 croissant for free, so she actually gets $10 + 1 = 11$ croissants for each box purchased. In total Dulce will get $140 \times 11 = 1540$ croissants.

(b) 3205 **boxes of croissants?**

Answer

35255

Solution

This time she would get $3205 \times 11 = 35255$ croissants.

Problem 2.24 Coach Robertson is setting up a huge soccer tournament and is inviting teams from around the whole country. He needs to set up hotel accommodations for all the teams attending. If every team will bring a total of 22 people and each room fits 2 people, how many rooms does he need to book if 365 teams come to the tournament?

Answer

4015

Solution

Since there are 365 teams coming and each team has 22 people, there will be 365×22 people coming to the tournament. Since each room fits 2 people, we just need to divide the number of people by 2 to find out how many rooms Coach need. This means he needs

$$365 \times 22 \div 2 = 365 \times 11 = 4015$$

rooms. (Note we simplified this computation by first calculating $22 \div 2 = 11$.)

Problem 2.25 Donna needs to buy some office supplies for her company. The supplier has a deal that for every 5 boxes she buys, she will get 1 extra box of the same supplies for free.

(a) How many pens did she get for free if she paid for 140 **boxes of** 10 **pens?**

280

Donna bought 140 boxes of pens, so she will get $140 \div 5$ boxes for free. Each box has 10 pens, so in the end she will get

$$140 \div 5 \times 10 = 140 \times 2 = 280$$

free pens.

(b) How many envelopes did she get for free if she paid for 1805 **boxes of** 100 **envelopes?**

36100

By buying 1805 boxes of envelopes, Donna gets $1805 \div 5$ boxes for free. Each box has 100 envelopes, so in the end she will get

$$1805 \div 5 \times 100 = 1805 \times 20 = 3610 \times 10 = 36100$$

free envelopes.

Problem 2.26 Ms. Pelegrino is grading some tests. There are 20 questions on each of the 5 pages of the test. Apparently her students did great on the test. So far she has the number of correct answers on each page of the test. Help her find out the total scores for the following two tests, based on the scores on each page.

(a) Correct answers per page: 20, 18, 19, 18 **and** 20.

95

Solution

Since this is almost a perfect score, it will be easier for Ms. Pelegrino to subtract the number of incorrect answers in each page from the maximum possible score of $20 \times 5 = 100$ points. Thus, the total score in this test is $20 \times 5 - 0 - 2 - 1 - 2 - 0 = 100 - 5 = 95$ points.

(b) Correct answers per page: 17, 20, 18, 15 and 17.

Answer

87

Solution

She can do the same for this one. The final score is $20 \times 5 - 3 - 0 - 2 - 5 - 3 = 100 - 13 = 87$ points.

Problem 2.27 **Sam is using colored square tiles to cover some canvases for his art class. The tiles are 1 inch long per side. How many tiles will he need for each of the canvases?**

(a) A square canvas that is 17 inches long per side.

Answer

289

Solution

Since it is a square canvas, Sam will need 17^2 tiles. Remember we learned a way to multiply "teen" numbers easily: $7 \times 7 = 49$, and $10 + 7 + 7 + 4 = 28$, so Sam will need 289 tiles for this canvas.

(b) A square canvas that is 35 inches long per side.

Answer

1225

Solution

This time Sam will need 35^2 tiles. This is a square that ends in 5, so we just need to multiply $3 \times 4 = 12$ and append 25. So, Sam will need 1225 tiles for this canvas.

(c) A rectangular canvas that is 12 inches long and 22 inches wide.

Answer

264

Solution

This canvas is not square, so we just need to multiply the length and height of the canvas, that is, Sam will need

$$12 \times 22 = 12 \times 2 \times 11 = 24 \times 11 = 264$$

square tiles for this canvas.

Problem 2.28 Compute the following. Before multiplying and dividing, make sure to group the numbers to simplify the calculation.

(a) $50 \times 25 \div 2$

Answer

625

Solution

$$\frac{\overset{25}{\cancel{50}} \times 25}{\cancel{2}} = 25 \times 25 = 625$$

(b) $350 \times 22 \div 7 \div 2$

Answer

550

Solution

$$\frac{350^{50} \times 22}{7 \times 2} = \frac{50 \times 22^{11}}{2} 50 \times 11 = 550$$

(c) $715 \div 44 \times 4$

Answer

65

Solution

$$\frac{715 \times 4}{44^{11}} = \frac{715}{11} = 65$$

Problem 2.29 Find the product of the following pairs of numbers. Notice that they start with the same digits and their last digits add up to 10.

(a) 67×63

Answer

4221

Solution

$6 \times (6+1) = 42$ and $7 \times 3 = 21$, so $67 \times 63 = 4221$.

(b) 24×26

Answer

624

Solution

$2 \times (2+1) = 6$ and $4 \times 6 = 24$, so $24 \times 26 = 624$.

(c) 83×87

Answer

7221

Solution

$8 \times (8+1) = 72$ and $3 \times 7 = 21$, so $83 \times 87 = 7221$.

Problem 2.30 **Find the product of the following pairs of numbers. Notice that they start with the same two digits and their last digits add up to 10.**

(a) 497×493

Answer

245021

Solution

$49 \times (49+1) = 2450$ and $7 \times 3 = 21$, so $497 \times 493 = 245021$.

(b) 311×319

Answer

99209

Solution

$31 \times (31+1) = 992$ and $1 \times 9 = 9$, so $311 \times 319 = 99209$.

(c) 783×787

Answer

616221

Solution

$78 \times (78+1) = 6162$ and $3 \times 7 = 21$, so $783 \times 787 = 616221$.

Problem 2.31 Group together the numbers that are multiplying and the numbers that are dividing. Write the expression using the horizontal line as division. Try to identify ways to simplify the numbers before you proceed to multiply and divide.

(a) $63 \times 12 \div 7 \div 9 \times 4$

$$\frac{\quad \times \quad \times \quad}{\times} =$$

Answer

48

Solution

$$\frac{63 \times 12 \times 4}{7 \times 9} = 12 \times 4 = 48$$

(b) $25 \times 3 \div 5 \div 2 \times 64$

$$\frac{\quad \times \quad \times \quad}{\times} =$$

Answer

480

Solution

$$\frac{25 \times 3 \times 64}{5 \times 2} = 15 \times 32 = 480$$

(c) $4 \div 3 \times 36 \div 10 \times 50$

$$\frac{\quad \times \quad \times \quad}{\times} =$$

Answer

240

Solution

$$\frac{4 \times 36 \times 50}{3 \times 10} = 4 \times 12 \times 5 = 4 \times 60 = 240$$

Problem 2.32 **With the aid of the given multiplication tables, find the product of the numbers.**

(a) 67×34

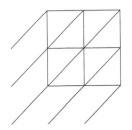

Answer

2278

(b) 93×82

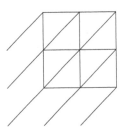

Answer

7626

(c) 74×29

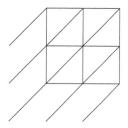

Answer

2146

Problem 2.33 With the aid of the given multiplication tables, find the product of the numbers.

(a) 124×394

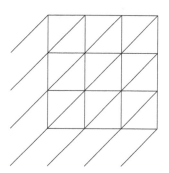

Answer

48856

(b) 567×291

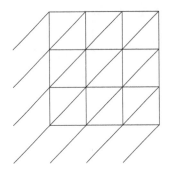

Answer

164997

(c) 696×313

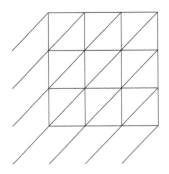

Answer

217848

Problem 2.34 **Use the rectangular multiplication tables to find the product of the given numbers.**

(a) 67×2369

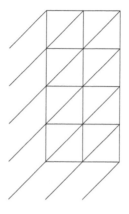

Answer

158723

(b) 6457×34

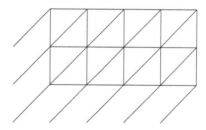

Answer

219538

(c) 3366×821

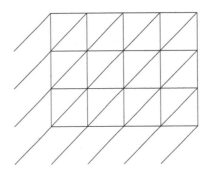

Answer

2763486

Problem 2.35 Use the rectangular multiplication tables to find the product of the given numbers.

(a) 13×481

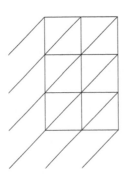

Answer

6523

(b) 761×34

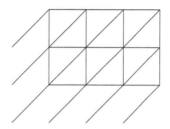

25874

(c) 49×273

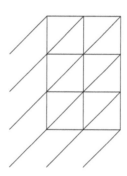

13377

3. Big and Small, Included

Problem 3.1 **The sum of the ages of Aiden and Brian is 12. Aiden's age is twice Brian's age. What are their current ages?**

Aiden: 8; Brian: 4

Solution

Let's pretend that both Aiden and Brian get a new toy car for their collection every year on their birthday. So, they have 12 cars in total, and Aiden has twice as many cars as Brian. Make small groups of $2 + 1 = 3$ cars, 2 from Aiden and 1 from Brian. We can make $12 \div 3 = 4$ groups like this, so Brian has 4 cars and Aiden has 8 cars. That is, Aiden is 8 years old and Brian is 4 years old.

Problem 3.2 **Justin got 1 out of every 7 questions wrong on a 105 point test. How many questions did Justin get correct?**

Answer

90

Solution

There are $105 \div 7 = 15$ groups of 7 questions in this exam. In each group of 7 questions, Justin got 1 incorrect and 6 correct questions, so in total he got $15 \times 6 = 90$ correct questions.

Problem 3.3 Adam gets \$1.75 for lunch every day. He bought a juice box for 25¢, and a burger for 5 times as much as the juice box. Does he have enough money left for an ice cream that costs 27¢?

Answer

No

Solution

In total Adam spent $25 \times (5 + 1) = 150$ cents buying the juice box and the burger. This means he has $175 - 150 = 25$ cents left, which is not enough for him to buy the ice cream.

Problem 3.4 Tom and Jerry went fishing. They caught 60 fish altogether. Tom caught 3 times as many fish as Jerry. How many fish did each catch?

Answer

Jerry: 15; Tom 45

Solution

Jerry caught less fish than Tom. For every 1 fish that Jerry caught, Tom caught 3, so Jerry caught

$$60 \div (3 + 1) = 15$$

fish and Tom caught $15 \times 3 = 45$ fish.

Problem 3.5 In Old McDonald's corn field, 2 out of 3 scarecrows dance the Macarena. If there are 120 scarecrows in the field, how many don't do this dance?

Answer

40

Solution

We can make groups of 3 scarecrows where 2 of them dance and 1 of them does not dance. Since we can make $120 \div 3 = 40$ of these groups, there are 40 scarecrows that do not dance the Macarena.

Problem 3.6 David's dad bought a digital camera from an online store. He paid the price of the camera plus shipping and handling, for a total of $270. The price of the camera is 8 times the cost of shipping and handling. What is the price of the digital camera?

Answer

$240

Solution

For every 1 dollar spent on shipping and handling, David's dad spent 8 dollars on the camera. So, he spent $280 \div (1+8) = 30$ dollars on shipping and handling, and $8 \times 30 = 240$ on the camera.

Problem 3.7 A country sent 108 athletes to the Olympic Games, among which the number of male athletes is twice the number of female athletes. How many male and female athletes are there?

Answer

Female: 36; Male: 72

Solution

There are less women on the Olympic team. For every 1 woman on the team there are 2 men, so there were $108 \div (2+1) = 36$ women and $108 - 36 = 72$ men on the Olympic team.

Problem 3.8 Aiden and Brandon collected 69 rare coins altogether. Aiden collected 2 times more coins than Brandon. How many coins did each of them collect?

Answer

Aiden: 46; Brandon: 23

Solution

Brandon has less coins than Aiden. For each coin that Brandon has, Aiden has 2, so Brandon collected $69 \div (2 + 1) = 23$ rare coins and Aiden collected $23 \times 2 = 46$ coins.

Problem 3.9 Suppose the height of an elf is 3 times the height of a hobbit. Also suppose that an elf is 60 inches taller than a hobbit. What is the height of each?

Answer

Hobbit: 30 inches; Elf: 90 inches

Solution

A hobbit is 60 shorter than an elf. For every 1 inch of the hobbit's height, the elf is $3 - 1 = 2$ inches taller. This means the hobbit is $60 \div 2 = 30$ inches tall and the elf is $30 \times 3 = 90$ inches tall.

Problem 3.10 A farm has some ducks and geese. There are 8 more ducks than geese. The number of ducks is 3 times the number of geese. How many ducks and how many geese are there in the farm?

Answer

Duck: 12; Geese: 4

Solution

There are more ducks than geese on the farm. For every 1 geese there are 3 ducks, so for every 1 geese there are 2 more ducks than geese. Since there are 8 more ducks than geese, there must be $8 \div 2 = 4$ geese and $4 \times 3 = 12$ ducks.

Problem 3.11 Your best friend is having a birthday party. Since his twin sisters' birthday is so close, their parents decided to have one party for the three of them at the same time. They ask you for help placing the candles on each of their cakes, but you can't remember how old each of them are becoming this year. They give you a total of 20 **candles, and you also know that this year the age of your friend will be** 3 **times the age his twin sisters. How many candles should you place in each cake?**

Answer

Friend's cake: 12 candles; Twin's cakes: 4 candles, each

Solution

You can make groups of $1 + 1 + 3 = 5$ candles. In each of this groups there will be 1 candle for each of the twins and 3 candles for your friend. Since you have 20 candles in total, each of the twins is $20 \div 5 = 4$ years old and your friend is $4 \times 3 = 12$ years old.

Problem 3.12 **You have two big buckets of water: one is red and the other one is blue. The red bucket contains 14 liters of water, and the blue bucket contains** 18 **liters of water. If you want bucket the red bucket to contain 3 times as much water as the blue bucket, how much water do you need to pour from the blue bucket into the red bucket?**

Answer

10 liters

Solution

We have in total $14 + 18 = 32$ liters of water. We want the red bucket to have 3 times as much water as the blue bucket, so for every 1 liter of water that we blue bucket has, we want the red bucket to have 3 liters. This means we want the blue bucket to have $32 \div (3 + 1) = 8$ liters of water. Right now the blue bucket has 18 liters of water, so we need to pour $18 - 8 = 10$ liters of water from the blue bucket to the red bucket.

Problem 3.13 **Monkey George had a basket of bananas. The total weight including the bananas and the basket is 32 pounds. Monkey George ate half of the bananas, and then the total weight including the bananas and the basket was** 17 **pounds. How many pounds of bananas were there originally? How heavy is the basket?**

Answer

Bananas: 30 pounds; Basket: 2 pounds

Solution

If we compare the total weight of the basket with bananas and the basket after eating half of the bananas, we can see that Monkey George ate $32 - 17 = 15$ pounds of bananas. This means there were $15 \times 2 = 30$ pounds of bananas in the basket, and so the weight of the basket is $32 - 30 = 2$ pounds.

Problem 3.14 **The sum of the ages of Adam and Bob is 14. Two years ago Adam's age was twice Bob's current age. What are their current ages?**

Answer

Adam: 10 ; Bob: 4

Solution

If Adam was 2 years younger, the sum of their ages would be $14 - 2 = 12$ and Adam would be twice as old as Bob. This means Bob is $12 \div (1 + 2) = 4$ years old and Adam would be $4 \times 2 = 8$ years old, so Adam is actually $8 + 2 = 10$ yeas old

Problem 3.15 **Sinbad, Popeye, and Captain Hook went on the sea to hunt for treasure. They found totally 1645 gold coins. Popeye found twice as many as Sinbad, and Captain Hook found twice as many as Popeye. How many gold coins did each of them find?**

Answer

Sinbad: 235; Popeye: 470; Captain Hook: 940

Solution

We can see Sinbad was the one that found the least amount of gold coins, followed by Popeye in second place and Captain Hook in first place. For every 1 coin that Sinbad found, Popeye found 2 and Captain Hook found $2 \times 2 = 4$. So, we can make groups of $1 + 2 + 4 = 7$ coins (1 from Sinbad, 2 from Popeye and 4 from Captain Hook). We can make $1645 \div 7 = 235$ of this groups. So, Sinbad found 235 coins, Popeye found $235 \times 2 = 470$ coins and Captain Hook found $470 \times 2 = 940$ coins.

Problem 3.16 **The sum of the heights of Mr. Giant and Mr. Super Giant is 45 feet. If Mr. Super Giant is 5 feet taller than 4 times the height of Mr. Giant, what is the height of each of them?**

Answer

Mr. Giant: 8 feet; Mr. Super Giant: 37 feet

Solution

Clearly Mr. Giant is shorter than Mr. Super Giant. If Mr. Super Giant was 5 feet shorter, he would be exactly 4 times as high as Mr. Giant and the sum of their heights would be $45 - 5 = 40$ feet. In this case, Mr. Giant would be $40 \div (4+1) = 8$ feet tall and Mr. Super Giant would be $8 \times 4 = 32$ feet tall. This means Mr. Giant is 8 feet tall and Mr. Super Giant is $32 + 5 = 37$ feet tall.

Problem 3.17 **There are 48 students in a class. If 3 more boys joined the class, the number of boys would be twice the number of girls. What is the current number of boys in the class?**

Answer

Girls: 17; Boys: 31

Solution

Let's pretend 3 more boys indeed join the class, so there are 2 times as many boys as girls and so $48 + 3 = 51$ students in total. In this case, since there are less girls than boys, there would be $51 \div (2+1) = 17$ girls and $17 \times 2 = 34$ boys. Since we assumed there were 3 more boys, we actually have 17 girls and $34 - 3 = 31$ boys.

Problem 3.18 **Jadean and Lucy have $32 in total. Before Jadean bought a $3 pen, he had 4 times as much money as Lucy did. How much money did each of them have originally?**

Answer

Lucy: $7; Jadean: $28

Solution

If Jadean hadn't bought that pen, they would have $32 + 3 = 35$ dollars in total and he would have 4 times as much money as Lucy. This means Lucy has $35 \div (4+1) = 7$ dollars and Jadean had $7 \times 4 = 28$ dollars.

Problem 3.19 You have $20, and your friend has $25. How much money should your friend give you so that you will have twice as much money as your friend?

Answer

$10

Solution

You and your friend have $20 + 25 = 45$ dollars in total. If the goal is for you to have 2 times as much money as your friend, in the end your friend should have $45 \div (2+1) = 15$ dollars. That means he should give you $25 - 15 = 10$ dollars.

Problem 3.20 A pen and a pencil together cost $2.10. If the price of a pen is 6 times the price of a pencil, what is the price of each?

Answer

Pencil: 30¢; Pen: $1.80

Solution

Since the cost of both one pen and one pencil is 210 cents and the pen is 6 times as expensive as the pencil, each pencil costs $210 \div (6+1) = 30$ cents, and a pen costs $30 \times 6 = 180$ cents, so $1.80.

Problem 3.21 Sam has $30 more than 3 times the amount of money Tom has, and Paul has $15 less than Tom. Altogether they have $240. How much money does Sam have?

Answer

$165

Solution

If Sam had 30 dollars less and Paul had 15 more dollars, they would have in total $240 - 30 + 15 = 225$ dollars, Paul and Tom would have the same amount of money , and Sam would have 3 times as much money as Tom. This way we could put their money in groups of $1 + 1 + 3 = 5$ dollars: 1 dollar for Paul, 1 dollar for Tom, and 3 dollars for Sam. This means each of Tom and Paul would have $225 \div 5 = 45$ dollars and Sam would have $45 \times 3 = 135$ dollars. So Sam actually has $135 + 30 - 165$ dollars.

Problem 3.22 **The sum of the ages of Lisa and Suzi is 24. Four years ago Lisa's age was three times Suzi's age. What are their current ages?**

Answer

Suzi: 8; Lisa: 16

Solution

Four years ago the sum of the ages of Lisa and Suzi was $24 - 4 - 4 = 16$ years. At that time, Lisa was 3 times as old as Suzi, so Suzi was $16 \div (3 + 1) = 4$ years old and Lisa was $4 \times 3 = 12$ years old. Thus, Lisa is $12 + 4 = 16$ years old, and Suzi is $4 + 4 = 8$ years old.

Problem 3.23 **Adam has 55 songs in his iPod, including pop songs and classical music. The number of pop songs is 4 times as many as the number of pieces of classical music. How many pop songs and how many pieces of classical music does he have?**

Answer

Classical: 11; Pop: 44

Solution

Since Adam has less pieces of classical music than pop songs in his iPod, and 55 songs in total, he has $55 \div (4 + 1) = 11$ classical music pieces and $11 \times 4 = 44$ pop songs in his iPod.

Problem 3.24 **Bob's dad was 25 years old when Bob was born. The age of Bob's dad last year was 3 times Bob's current age. How old are Bob and his dad this year?**

Answer

Bob: 12; Bob's dad: 37

Solution

Bob is 25 younger than his dad. The age of his dad last year and his current age are then $25 - 1 = 24$ years apart. His dad's age last year is 3 times his current age, so for each 1 year in Bob's age there are 2 more years in his dad's age last year. This means Bob is $24 \div 2 = 12$ years old and his dad is $12 \times 3 + 1 = 37$ years old.

Problem 3.25 **Old McDonald and Old Wendy went to the market to sell apples. They originally had the number of pounds of apples. Old McDonald sold 11 pounds of apples, and Old Wendy sold 29 pounds. Now Old McDonalds has 3 times as many pounds of apples as Old Wendy. How many pounds of apples do each of them have now?**

Answer

Old Wendy: 9 pounds; Old McDonald: 27 pounds

Solution

Old Wendy sold $29 - 11 = 18$ more pounds of apples than Old McDonald. Since they had the same amount of apples at first, this means now the difference between the amount of apples they have is 18 pounds and Old Wendy has less apples than Old McDonald. Since now Old McDonald has 3 times as many pounds of apples as Old Wendy, Old McDonald has $3 - 1 = 2$ more pounds of apples for every 1 pound of apples that Old Wendy has. Thus, Old Wendy has $18 \div 2 = 9$ pounds of apples and Old McDonald has $3 \times 9 = 27$ pounds of apples.

Problem 3.26 **A farm produced $1,600$ pounds of fruit, including bananas and oranges. They produced 100 more pounds of bananas than 3 times the number of pounds of oranges they produced. How many pounds of bananas and how many pounds of oranges did they produce?**

Answer

Oranges: 375 pounds; Bananas: 1225 pounds

Solution

If the farm had produced 100 less pounds of bananas, they would have produced exactly 3 times as many pounds of bananas as oranges, and they would have $1600 - 100 = 1500$ pounds of fruit in total. This means they produced $1500 \div (3 + 1) = 375$ pounds of oranges and $1600 - 375 = 1225$ pounds of bananas.

Problem 3.27 **There are boys and girls in the classroom. Had there been** 10 **fewer boys, the numbers of boys and girls would have been equal. Had there been** 10 **fewer girls, the number of boys would have been twice the number of girls. How many boys, and how many girls are there?**

Answer

Girls: 30; Boys: 40

Solution

There are 10 more boys than girls in the classroom. If there were 10 less girls in the classroom, there would be $10 + 10 = 20$ more boys in the classroom, and there would be 2 times as many boys as girls. So, there would be 1 more boy than girls for every 1 girl in the classroom. Therefore there would be 20 girls in the classroom and $20 + 20 = 40$ boys. Since we assumed there were 10 less girls in the classroom, we actually have 40 boys and $20 + 10 = 30$ girls.

Problem 3.28 **Gold Toothed Brendan and Peg Legged Samuel went to hunt for treasures. Gold Toothed Brendan found** 29 **more diamonds than Peg Legged Samuel. Also, Gold Toothed Brendan found** 1 **more diamond than** 3 **times the number of diamonds that Peg Legged Samuel found. How many diamonds did each of them find?**

Answer

Peg Legged Samuel: 14; Gold Toothed Brendan: 43

Solution

If Gold Toothed Brendan had found 1 less diamond, he would have found exactly 3 times as many diamonds than Peg Legged Samuel and he would have found $29 - 1 = 28$ more diamonds than Peg Legged Samuel. So, for every 1 diamond found by Peg Legged Samuel, Gold Toothed Brendan would have found 2 more. Thus, Peg Legged Samuel found $28 \div 2 = 14$ diamonds and Gold Toothed Brendan found $14 \times 3 + 1 = 43$ diamonds.

Problem 3.29 **Clark Kent and Neo are comparing their flying speeds. The speed of Clark Kent is 540 mph faster than that of Neo. The speed of Clark Kent is 90**

mph less than 4 times the speed of Neo. What is the flying speed of each of them?

Answer

Clark Kent: 750 mph; Neo: 210 mph

Solution

If Clark Kent was 90 mph faster, the difference of their speeds would be $540 + 90 = 630$ mph and he would be 4 times as fast as Neo. In this case, for every 1 mph in Neo's speed, the speed of Clark Kent is $4 - 1 = 3$ mph faster than Neo's. This means Neo's speed is $630 \div 3 = 210$ mph and Clark Kent's is $210 + 540 = 750$ mph.

Problem 3.30 Bugs and Tweety were bored so they decided to throw pebbles into the lake. Originally they had the same number of pebbles. Bugs threw 7 pebbles and Tweety threw 19, then Bug had 3 times as many remaining pebbles as Tweety. How many pebbles do each of them have now?

Answer

Bugs: 18; Tweety: 6

Solution

Since they started with the same number of pebbles, now Tweety has $19 - 7 = 12$ less pebbles than Bugs. For every 1 pebble that Tweety has, Bugs has $3 - 1 = 2$ more pebbles than Tweety. So, Tweety has $12 \div 2 = 6$ pebbles and Bugs has $6 \times 3 = 18$ pebbles.

Problem 3.31 Mickey, Donald and Goofy went for a bicycle ride. If Mickey rode 6 miles more than Donald, Goofy rode 22 miles more than Donald, and Goofy rode twice as far as Mickey, how far did each of them ride?

Answer

Mickey: 16 miles; Donald: 10 miles; Goofy 32 miles

Solution

Goofy rode 22 more miles than Donald and Donald rode 6 less miles than Mickey, so Goofy rode $22 - 6 = 16$ more miles than Mickey. Since Goofy rode 2 times as many miles as Mickey, for every 1 mile that Mickey rode, Goofy rode $2 - 1 = 1$ more mile than Mickey. This means Mickey rode 16 miles and Goofy rode $16 \times 2 = 32$ miles.

Then Donald rode for $32 - 22 = 10$ miles.

Problem 3.32 A baseball game is being played. Some people are watching at the stadium and some are watching on TV at home. The number of people who watch on TV is 480 more than the number of people in the stadium. If 50 people at the stadium went home and watched the game on TV, the number of people who watch on TV would be 5 times the number of the number of people in the stadium. How many people are watching the game in total?

Answer

870

Solution

When 50 people at the stadium go home and keep watching, there are $480 + 50 + 50 = 580$ more people watching at home than at the stadium. For every 1 person in the stadium there are 4 more people watching at home than in the stadium. So, there are $580 \div 4 = 145$ people in the stadium and $145 + 580 = 725$ people watching at home. That is, $725 + 145 = 870$ people in total.

Problem 3.33 Harry and Hermione are making magic potions. If Harry added 8 ounces of potion into his cauldron, he would have the same amount of potion as Hermione. If Harry poured 3 ounces of potion from his cauldron into Hermione's, then Hermione would have 3 times as much potion as Harry. How much potion do each of them have originally?

Answer

Harry: 10 ounces; Hermione: 18 ounces

Solution

Harry has 8 ounces of potion less than Hermione on his cauldron. When he pours 3 ounces of his potion into Hermione's cauldron, Hermione has $8 + 3 + 3 = 14$ more ounces of potion in her cauldron and also has 3 times as much potion as Harry. So, for every 1 ounce of potion in Harry's cauldron, there are 2 more ounces in Hermione's cauldron than in Harry's. This means Harry has now $14 \div 2 = 7$ ounces of potion in his cauldron and Hermione has $7 \times 3 = 21$ ounces. Since we assumed Harry poured 3 ounces from his cauldron into Hermione's, we have that Harry actually has $7 + 3 = 10$ ounces of potion and Hermione has $21 - 3 = 18$ ounces.

Problem 3.34 A room is rectangular in shape, and its length is 4 times its width. Given that the length is 36 feet longer than the width, find the area of the room in square feet.

Answer

576 square feet

Solution

The width is smaller than the length, and for every 1 foot in the width, there are $4-1=3$ feet in the length. Therefore, the width is $36 \div 3 = 12$ feet and the length is $12 \times 4 = 48$ feet. The area of the rectangle is then $12 \times 48 = 576$ square feet.

Problem 3.35 You have \$56, and your friend has \$40. How much money should you give to your friend so that you two have the *same* amount of money?

Answer

\$8

Solution

You and your friend have $56 + 40 = 96$ dollars in total. You want each to have the same amount of money, so each of you should have in the end $96 \div 2 = 48$ dollars. Thus, you should give your friend $56 - 48 = 8$ dollars.

4. Come Together, Leave Apart

Problem 4.1 There are 65 oranges in a grocery store arranged in two piles. If the first pile has 5 more oranges than the second pile, how many oranges does each pile have?

Answer

30 and 35

Solution

The smaller pile will have

$$(65 - 5) \div 2 = 30$$

oranges. The big pile will have

$$30 + 5 = 35$$

oranges.

Problem 4.2 When Ian walks on stilts he is 17 feet tall. If he stands on the ground next to his stilts, the distance from his head to the top of the stilts is 5 feet. How tall is Ian?

Answer

6 feet

Solution

Ian is shorter than his stilts, so he is

$$(17-5) \div 2 = 6$$

feet tall.

Problem 4.3 Alek and Abby are collecting toy cars. Together they have 35 toy cars and Abby has 7 less than Alek. How many toy cars does Alek have?

Answer

21

Solution

Alek has more cars than Abby, so he has

$$(35+7) \div 2 = 21$$

cars.

Problem 4.4 Patrick and Joe ate all the cookies that their mom had just made. If her mom made 22 cookies and Joe ate 10 more cookies than Patrick. How many cookies did each of them eat?

Answer

Patrick: 6; Joe: 16

Solution

Patrick ate less cookies than Joe, so he ate

$$(22-10) \div 2 = 6$$

cookies and Joe ate $6 + 10 = 16$ cookies.

Problem 4.5 **Tim is 5 years old, and Allison is 3 years old. Their mom says she will take them to China when their combined age is 20. How old are they when they go to China?**

Answer

Tim: 11; Allison: 9

Solution

The difference of their ages is $5 - 3 = 2$. Their ages will add up to 20 when Allison is

$$(20 - 2) \div 2 = 9$$

years old and Tim is

$$9 + 2 = 11$$

years old.

Problem 4.6 **Josh had 10 chocolate galleons that he shared with his friend Mike. Mike says that Josh ate 3 more galleons than him. Is that possible? How many did each of them eat?**

Answer

Mike: 3 and a half; Josh: 6 and a half

Solution

Mike ate less chocolate than Josh, so he ate

$$(10 - 3) \div 2 = 3.5$$

chocolate galleons and Josh ate $10 - 3.5 = 6.5$ chocolate galleons.

Problem 4.7 **One pig and one duck weigh 98 pounds. One pig and one turkey weigh 123 pounds. One duck and one turkey weigh 37 pounds. What is the weight of the turkey?**

Answer

31

Solution

If we add up the weights of each of the pairs we have added up the weight of all three animals twice, so the sum of their weights (only one time each) is

$$(98 + 123 + 37) \div 2 = 129$$

pounds. Since the weight of the pig and the duck is 98 pounds, then the weight of the turkey is $129 - 98 = 31$ pounds.

Problem 4.8 David likes collecting his bus tickets. Last week he took the bus 6 times less than this week and he gathered 34 bus tickets altogether. How many times did he ride the bus this week?

Answer

20

Solution

This week David took the bus more times than last week, so he took the bus

$$(34 + 6) \div 2 = 20$$

times.

Problem 4.9 Quincey bought two boxes of pencils. One of the boxes was missing 4 pencils. If she has 62 pencils in total, how many pencils should each box have had?

Answer

33

Solution

Each box of pencils should have

$$(62 + 4) \div 2 = 33$$

pencils.

Problem 4.10 There were two big stacks of boxes in a storage unit your dad works in. When your dad puts 3 boxes of the big stack into the small stack, they

have the same height. If there are 16 **boxes in total, how many boxes are there in each stack originally?**

Answer

5 and 11

Solution

There are $3 + 3 = 6$ more boxes in the bigger stack than the smaller stack, so the bigger stack of boxes originally had

$$(6 + 16) \div 2 = 11$$

boxes, and the smaller stack had $16 - 11 = 5$ boxes.

Problem 4.11 **Four years ago, Curtis was** 36 **years older than Cassie. If this year their ages add up to** 100 **years, how old is Curtis?**

Answer

68 years old

Solution

The difference of their ages will not change over time, so this year Curtis is

$$(100 + 36) \div 2 = 68$$

years old.

Problem 4.12 **Olive and Bluto collect stamps. They have** 28 **stamps in total. If Olive gives** 6 **stamps to Bluto, she still has** 2 **more stamps than Bluto. How many stamps do each of them have originally?**

Answer

Olive: 21; Bluto: 7

Solution

Olive has $6 + 6 + 2 = 14$ more stamps than Bluto, and together they have 28, so Bluto has

$$(28 - 14) \div 2 = 7$$

stamps and Olive has $28 - 7 = 21$ stamps.

Problem 4.13 Again, the grocery store has two piles of oranges. There are 100 oranges in total. After 20 oranges are sold from the first pile, the two piles have the same number of oranges. How many oranges did each pile originally have?

Answer

40 and 60

Solution

The bigger pile had 20 more oranges than the smaller pile. Then the bigger pile had

$$(100 + 20) \div 2 = 60$$

oranges, and the smaller pile had

$$100 - 60 = 40$$

oranges.

Problem 4.14 Kyle and Lisa each bought some candies. If Kyle gave 4 candies to Lisa, they would both have the same number of candies. They pooled their candies together and counted, and there were 24 in total. How many candies did each of them originally have?

Answer

Kyle: 16; Lisa: 8

Solution

Kyle has $4 + 4 = 8$ more candies than Lisa. Kyle has more candies, so he must have

$$(24 + 8) \div 2 = 16$$

pieces of candy and Lisa $16 - 8 = 8$ pieces of candy.

Problem 4.15 Patrick had three exams: history, math, and literature, and got a total score of 250 points. The score he got in the literature exam was 20 points less than the score he got in math, and was 10 points more than the score he got in the history exam. How many points did he receive in each exam?

Answer

Literature: 80; History: 70; Math: 100

Solution

If Patrick had gotten 20 points less in the math test and 10 points more in the history score, he would have had $250 - 20 + 10 = 240$ points in total and he would have had the same score in all exams. This means he would have had a score of $240 \div 3 = 80$ in all exams. Thus, he actually got $80 + 20 = 100$ points in math, 80 points in literature and $80 - 10 = 70$ points in history.

Problem 4.16 Minsung and Patrick were playing Monopoly with some other friends. The number of properties that Patrick and Minsung have altogether is 24. If Patrick has 6 more properties than Minsung, how many properties does Minsung have?

Answer

9

Solution

Minsung has less properties than Patrick, so he has

$$(24 - 6) \div 2 = 9$$

properties.

Problem 4.17 Terrence has a small workshop where he builds chairs on his spare time. His son sometimes comes to help. When they work together they build 32 chairs in one weekend (2 days). If Terrence's son can build 4 less chairs than him in one day, how many chairs did each of them build over one weekend?

Answer

Terrence: 20; Terrance's son: 12

Solution

Terrence's son can build $4 + 4 = 8$ less chairs than him in a weekend. This means Terrence built

$$(32 + 8) \div 2 = 20$$

chairs and his son built $32 + 20 = 12$ chairs over one weekend.

Problem 4.18 **Adam, Bob and Chris have $875 altogether. Adam has $250 more than Bob, and Bob has $125 more than Chris. How much money do each of them have?**

Answer

Chris: 125; Bob: 250; Adam: 500

Solution

Adam has $250 + 125 = 375$ dollars more than Chris and Bob has 125 more than Chris. If Adam had 275 dollars less and Bob had 125 dollars less, they would all have the same amount of money, and they would have $875 - 375 - 125 = 375$ dollars altogether. This means each of them would have $375 \div 3 = 125$ dollars. So, Chris has 125 dollars, Bob has $125 + 125 = 250$ dollars and Adam has $125 + 275 = 500$ dollars.

Problem 4.19 **We planted 121 different plants in the school's garden. Unfortunately some of the plants got eaten by some squirrels. The number of plants that survived is 61 more than the number of plants eaten by the squirrels. How many of the plants did the squirrels eat?**

Answer

30

Solution

Since the number of plants that survived is greater than the number of plants that were eaten by squirrels,

$$(121 - 61) \div 2 = 30$$

were eaten by squirrels.

Problem 4.20 **In a school there are totally 108 students in the third, fourth, and fifth grades. The third grade has 11 fewer students than the fourth grade, and the fourth grade has 16 more students than the fifth grade. How many students are there in each grade?**

Answer

3rd grade: 34; 4th grade: 45; 5th grade: 29

Solution

If the third grade had 11 students more and the fifth grade had 16 students more, all of the three classes would have the same number of students and they would have $108 + 11 + 16 = 135$ students in total. Thus, the fourth grade has $135 \div 3 = 45$ students, the third grade has $45 - 11 = 34$ students, and the fifth grade has $45 - 16 = 29$ students.

Problem 4.21 Hercules bought 5 swords and 2 spears for 304 drachmas (drachma is the currency of Greece). A spear cost 9 drachmas less than a sword. How much does a sword cost?

Answer

Sword: 46 drachmas; Spear: 37 drachmas

Solution

If the spears were 9 drachmas more expensive, swords and spears would cost the same, and Hercules would have spent $304 + 2 \times 9 = 322$ drachmas buying all 7 items. In that case, each item would have cost $322 \div 7 = 46$ drachmas. Thus, each sword costs 46 drachmas and each spear costs $46 - 9 = 37$ drachmas.

Problem 4.22 Frank took three quizzes, with a total score of 148 points. His quiz A score is 21 points higher than his quiz B score, and his quiz C score is 32 points lower than his quiz A score. What is his score for each of the quizzes?

Answer

Quiz A: 67; Quiz B: 46; Quiz C: 35

Solution

If Frank had gotten 21 points more on his quiz B and 32 more on his quiz C, he would have had

$$148 + 21 + 32 = 201$$

points in total, while getting the same score in each quiz. This means he got $201 \div 3 = 67$ points in quiz A, $67 - 21 = 46$ points on quiz B and $67 - 32 = 35$ points on quiz C.

Problem 4.23 Old Sealegs Austin, Landlubber Matthew, and Cap'n Poopdeck went to search for treasure. They found 130 gold coins in total. Old Sealegs Austin found 7 more gold coins than Landlubber Matthew, and if Old Sealegs Austin and Landlubber Matthew put their gold coins together they would have 24 more than Cap'n Poopdeck's. How many gold coins did each of them find?

Answer

Old Sealegs Austin: 42; Landlubber Matthew: 35; Cap'n Poopdeck: 53

Solution

Let's find first how many coins did Landlubber Matthew and Old Sealegs Austin found together. Since they found 24 more coins than Cap'n Poopdeck and all three of them found 130 coins in total, this means that Captain Poopdeck found

$$(130 - 24) \div 2 = 53$$

coins, and they found $130 - 53 = 77$ coins together. Now that we know how many coins they got together, knowing that Old Sealegs Austin found 7 more than Landlubber Matthew, we have that Old Sealegs Austin found

$$(77 + 7) \div 2 = 42$$

coins and Landlubber Matthew found $77 - 42 = 35$ coins.

Problem 4.24 Of the 80 levels I completed on my video game, it took two tries to complete some levels and only one try to complete other levels. The number of levels which only took one try is 20 more than the number of levels which took two tries. How many levels only took one try?

Answer

50

Solution

The number of levels that took only one try is bigger than the ones that took two tries, so there were

$$(80 + 20) \div 2 = 50$$

levels that took only one try.

Problem 4.25 **I tried to squirt Dad with a hose** 101 **times last summer. I missed** 31 **more times than the number of times I got him wet. How many times did I miss trying to squirt Dad?**

66

The number of times I missed is greater than the number of times I got him wet. Thus, I missed

$$(101 + 31) \div 2 = 66$$

times.

Problem 4.26 **Myles, Mya and Myron entered a team competition where they each had to run** 400 **meters and add up their times to get their team score. Myles was** 10 **seconds faster than Myron, and Mya was** 4 **seconds faster than Myron. If their total time was** 181 **seconds, how long did it take Myron to complete the course?**

65 seconds

If Myles had been 10 seconds slower and Mya had been 4 seconds slower, they would all have had the same time as Myron and their team score would have been $181 + 10 + 4 = 195$ seconds. This way each of them would have taken $195 \div 3 = 65$ seconds to complete the 400 m run. Thus Myron completed the course in 65 seconds.

Problem 4.27 **The twins Walker and Cooper are coming to a Halloween party dressed up as a gigantic ghost (so, Walker will be standing on Cooper's shoulders and then they will cover themselves with a huge blanket). They were so tall that Walker hit his head with a lamp that was** 11 **feet high. If their heads are** 1 **foot long, how tall are the twins?**

6 feet

Solution

The "ghost" is 11 feet high, which means that the height of one twin and the height of a twin with no head add up to 11 feet. Since the heads are 1 foot long, the difference in height between a whole body and a body with no head is 1 foot. Therefore the twins are

$$(11+1) \div 2 = 6$$

feet tall.

Problem 4.28 Cameron, Mandy and Taylor decided to put all their savings together to buy the new video game console that all the kids were playing. They had enough money to buy the video game console that cost $300 and an extra controller that cost $20. If Taylor had $20 less than Mandy, and Mandy had $50 more than Cameron, how much money did each of them have?

Answer

Cameron: $80; Mandy: $130; Taylor: $110

Solution

Together they had $300 + 20 = 320$ dollars. If Taylor had had 20 dollars more and Cameron had had 50 dollars more, they would have had $320 + 30 + 50 = 390$ dollars in total and they would all have had the same amount of money. This means Mandy had $390 \div 3 = 130$ dollars, Cameron had $130 - 50 = 80$ dollars and Taylor had $130 - 20 = 110$ dollars.

Problem 4.29 Wilson watched the movie Cast Away and got obsessed with Wilson, the volleyball, because they have the same name. He visited some sports stores and every time he bought more volleyballs than in the previous store. He bought in total 19 volleyballs in three different sports stores. In the second store he bought 2 more balls than in the first store and in the third store he bought 3 more than in the second store. How many balls did he buy in the first store?

Answer

4 volleyballs

Solution

Wilson bought 2 more balls in the second store, and $2 + 3 = 5$ more balls in the third store than in the first store. So, if he had bought 2 less balls in the second store and 5 less volleyballs in the third store, he would have bought $19 - 2 - 5 = 12$ balls in total and he would have bought the same number of volleyballs in each store. This way we know Wilson bought $12 \div 3 = 4$ volleyballs in the first store.

Problem 4.30 **Cody likes separating his M&M's by color before eating them. He has now only blue and green M&M's left. Before he ate 3 blue M&M's and 7 green M&M's, he had 50 M&M's. Now he has 2 more green M&M's than blue M&M's. How many green M&M's does he have left?**

Answer

21

Solution

After eating 3 blue and 7 green M&M's, he has $50 - 3 - 7 = 40$ M&M's left. Of those, he has 2 more green than blue, so he has

$$(40 + 2) \div 2 = 21$$

green M&M's and $21 - 2 = 19$ blue M&M's.

Problem 4.31 **Rosemary loves cooking with fresh herbs so she has some pots with thyme and rosemary, which is her favorite, in her backyard. If she had 2 more thyme pots and 4 less rosemary pots, she would still have 10 more rosemary pots than thyme pots. If she has 24 pots in total, how many pots of each plant does she have?**

Answer

Thyme: 4; Rosemary: 20

Solution

Let's pretend she indeed has 2 more thyme pots and 4 less rosemary pots. This way she has $24 + 2 - 4 = 22$ pots in total and she still has 10 more rosemary pots than thyme pots. This means she would have

$$(22 + 10) \div 2 = 16$$

rosemary pots and $16 - 10 = 6$ thyme pots. Since we pretended we had 2 more thyme pots and 4 less rosemary pots, Rosemary actually has $16 + 4 = 20$ rosemary pots and $6 - 2 = 4$ thyme pots.

Problem 4.32 Claude needed to weigh his dog Ty, but his other two dogs, Jy and Xy, were giving him problems. Every time he placed Ty on the scale at least one of the other dogs stepped on it too. If Ty and Jy weighed 103 pounds, Ty and Xy weighed 112 pounds, and the three of them together weighed 175 pounds, how heavy is Ty?

Answer

40 pounds

Solution

If Ty had had an identical twin and the four of them had stepped on the scale at the same time, their weight would be $103 + 112 = 215$ pounds. Since the three of them together weigh 175 pounds, Ty alone weighs $215 - 175 = 40$ pounds.

Problem 4.33 Two hungry ant-eaters were feeding from a small anthill. If one of the ant-eaters eats 150 more ants than the other and they ate a total of 2300 ants, how many did each of them eat?

Answer

1075 and 1225

Solution

The less hungry ant eater ate

$$(2300 - 150) \div 2 = 1075$$

ants and the other anteater ate $2300 - 1075 = 1225$ ants.

Problem 4.34 Mom and I went to the store to buy some utensils for a big dinner party we will be having over the weekend. We already had some so we didn't need to buy whole sets. We bought spoons, forks and knives, 70 in total. If we bought 10 more spoons than forks and 5 less spoons than knives, how many of each utensil did we buy?

Answer

Spoons:25, Forks:15; Knives: 30

Solution

If we had bought 10 more forks and 5 less knives, we would have bought the same number of each utensil having $70 + 10 - 5 = 75$ in total, so $75 \div 3 = 25$ of each. This means we actually bought 25 spoons, $25 - 10 = 15$ forks and $25 + 5 = 30$ knives.

Problem 4.35 **Rita needed to buy some cat food. Her cat, Melrose, is picky when she eats so she needs to buy different kinds of food. In total Rita bought 38 cans of food. She bought 4 more cans of "Savory Shreds" than "Classic Paté", and 10 more cans of "Prime Filets" than "Classic Paté". How many cans of each kind did she buy?**

Answer

Savory Shreds: 12; Classic Paté: 8; Prime Filets:18

Solution

Had Rita bought 4 less cans of "Savory Shreds" and 10 less cans of "Prime Filets" she would have bought $38 - 4 - 10 = 24$ cans in total, so $24 \div 3 = 8$ of each. This means she bought $8 + 4 = 12$ cans of "Savory Shreds", 8 cans of "Classic Paté" and $8 + 10 = 18$ cans of "Prime Filets".

5. Counting Without Fingers

Problem 5.1 **Find the number of 2-digit positive integers whose digits sum a total of 7.**

Answer

7

Solution

A two digit number cannot start with 0, and if we use a digit bigger than 7 the sum of its digits would be more than 7. For each digit we choose for the first digit of our number, we can always find a digit to put on the second digit so that the sum of both is 7, so there will be exactly 7 2-digit numbers whose digits add to 7. The numbers are: 16, 25, 34, 43, 52, 61, and 70.

Problem 5.2 **If 5 people are in a room and everyone shakes everyone else's hand once, how many handshakes were exchanged in the room?**

Answer

10

Solution

The first person to enter the room will not shake hands with anybody (yet). The second person to enter the room will shake hands with 1 person. The third will do with 2 people. Fourth with 3 people. The last person will do with 4 people. So, in total there were $1 + 2 + 3 + 4 = 10$ handshakes in the room.

Problem 5.3 **How many 4-digit numbers can you form with the digits 3, 7, 8 and 9 if you can use each digit only once?**

Answer

24

Solution

We have 4 options to choose from for the first digit. Once we choose the first digit we only have 3 options left for the second digit. Then we only have 2 options left for the third digit. For the last digit we have only 1 choice left. This means we can form $4 \times 3 \times 2 \times 1 = 24$ different numbers.

Problem 5.4 **How many 3-digit numbers can you form with the digits 1, 5 and 7? You can repeat digits.**

Answer

27

Solution

For each of the digits we will have the same 3 choices since we can repeat digits. This means we can make $3 \times 3 \times 3 = 27$ different numbers using those three digits.

Problem 5.5 **In how many ways can you choose what jacket and hat to wear if you have 4 jackets, 7 baseball caps, and 2 beanie hats? (You will only wear one hat at a time!)**

Answer

36

Solution

Assuming you always wear one jacket and one hat, you have 4 jackets to choose from and $7 + 2 = 9$ hats to choose from. Then you have $4 \times 9 = 36$ different ways of choosing what jacket and hat to wear.

Problem 5.6 When writing the integers 120 through 180, how many times is the digit 3 written?

Answer

16

Solution

Every 10 consecutive numbers we write the digit 3 one time in the ones digit. This happens 6 times between the numbers 120 and 180. We will also write the digit 3 in the tens place in the numbers 130 through 139, so there are 10 more. Thus, in total we will write the digit 3 $10 + 6 = 16$ times.

Problem 5.7 Jamie flips a coin 3 times. How many possible ways can Jamie get exactly 2 heads?

Answer

3

Solution

If Jamie can flip a coin 3 times, the possible ways that Jamie can flip 3 times are:

$$HHH, HHT, HTH, THH, TTH, THT, HTT, TTT.$$

Of these 8 possible ways, there are 3 ways that Jamie can get exactly 2 heads (HHT, HTH and THH).

Problem 5.8 How many different sums can you get adding two or more of the numbers 1, 3, 4 and 7?

Answer

9

Solution

Adding exactly two numbers we can get: $1+3=4$, $1+4=5$, $1+7=8$, $3+4=7$, $3+7=10$ and $4+7=11$. Adding exactly three numbers we can get $1+3+4=8$, $1+3+7=11$, $1+4+7=12$ and $3+4+7=14$. Adding all of them we can get $1+3+4+7=15$. So we can get 4, 5, 7, 8, 10, 11, 12, 14 and 15, that is, 7 sums. Note that we could not just count how many ways to choose two numbers to add (6 ways), which three numbers to add (4 ways) and add them all (1 way), since some of the sums are repeated.

Problem 5.9 **How many different 4-digit numbers can be formed by rearranging the four digits in** 2004?

Answer

6

Solution

Note that the number must start with a digit that is not 0, so the first digit should be either 2 or 4. Once we chose which digit we want to use first we just need to decide where we want to place the 0's and the other digit. We can do this in 3 different ways. So, in total there are $2 \times 3 = 6$ different numbers we can get by rearranging the digits of 2004.

Problem 5.10 **Of the** 30 **letters shown below, how many of them are** *O*s?

IOOOOOOOOOOOOOOOOOOOOOOOOOOOOI

Answer

28

Solution

Since the problem is telling us that there are 30 letters and we can see that there are only 2 that are not *O*s, there must be $30-2=28$ *O*s.

Problem 5.11 There are 3 different kinds of chili, 2 kinds of tomatoes, 3 kinds of onions and 2 kinds of avocado at the grocery store. You want to make some guacamole choosing one of each kind of these ingredients. In how many different ways can you choose the ingredients?

Answer

36

Solution

Since we want to choose exactly one of each kind of ingredient, we just need to multiply together the number of choices we have for each of them. So, in total we will have $3 \times 2 \times 3 \times 2 = 36$ different ways of making guacamole.

Problem 5.12 Thomas has a pen that he can only use for writing numbers and will run out of ink right after he writes the digit 7 a total of 10 times. He wants to write the counting numbers starting from 11. What is the last number Thomas will be able to write?

Answer

73

Solution

Thomas would need to write one time the digit 7 in the ones place every 10 consecutive numbers. So starting from 11, when he writes up until the number 69, he has written the digit 7, a total of 6 times (17, 27, 37, 47, 57 and 67) so he has only 4 more digit 7s available o write. Then he will also need to write the digit 7 in the tens: 70, 71, 72 and 73. So, the last number he can write is 73.

Problem 5.13 Thomas bought a new and improved pen that allows him to write up to 20 times the digit 7. This time he starts writing the counting numbers from 100. What is the last number he will be able to write?

Answer

197

Solution

Every 100 consecutive numbers the digit 7 is used 10 times in the ones place and 10 times in the tens place, so exactly the 20 times Thomas has available with this new pen. This means that the last number he will be able to write will be the last number that has a 7 between 100 and 200, that is the number 197.

Problem 5.14 **How many 4-digit even numbers can you form with without repeating digits and without using the digit 0?**

Answer

1344

Solution

Since we want the number to be even, that means we want the number to end in 2, 4, 6 or 8, so there are 4 possible choices for the last digit. Since we do not want to repeat digits, we cannot use that last digit in any of the other three digits of our number. So, for the first, second and third digits we have $9 - 1 = 8$, $8 - 1 = 7$ and $7 - 1 = 6$ choices, respectively. Thus, in total we can form $4 \times 8 \times 7 \times 6 = 1344$ different 4-digit even numbers without using the digit 0.

Problem 5.15 **Ricky has to choose 8 of his 10 favorite toys to bring on a long road trip. In how many ways can he choose what toys to bring?**

Answer

45

Solution

Note that choosing 8 of his 10 favorite toys to bring is the same as choosing 2 of his favorite toys to leave behind (and that is easier to count!). Since he has 10 toys, he has 10 choices for the first toy he leaves behind and 9 choices for the second toy he leaves behind. However, the order in which he chooses the toys that he wants to leave behind does not matter (it is the same if he leaves "his red car and his teddy bear" than if he leaves "his teddy bear and his red car"), so in total he has

$$10 \times 9 \div 2 = 45$$

ways of choosing what toys to leave behind, and thus he also has 45 ways of choosing which toys to bring along to the trip.

Problem 5.16 You are helping your Mom to make a fruit arrangement. How many different arrangements can you make if you must choose one from 3 kinds of pears, one from 6 kinds of berries, one from 4 kinds of apples, and one from 7 kinds of nuts?

Answer

504

Solution

Since we must choose exactly one of each kind, we just need to multiply the number of choices we have for each type of fruit. So you have $3 \times 6 \times 4 \times 7 = 504$ different ways to choose which fruits to use in the arrangement.

Problem 5.17 Three friends have a total of 6 identical pencils, and each one has at least one pencil. In how many ways can this happen?

Answer

10

Solution

Since we know that each of the three friends has at least one pencil we can go ahead and give one pencil to each of them. This means we need to figure out how to give away the other $6 - 3 = 3$ pencils we have left. We could give all three pencils to one of the friends (we can do this in 3 ways, just choose which friend gets them), or we could give 2 pencils to one of them and 1 pencil to different friend (we have 3 choices for who to give 2 pencils, and then 2 choices for who to give 1 pencil, so there are $3 \times 2 = 6$ ways to do this), or we could give exactly one pencil to each of the friends (there is only one way to do this). So in total we have $3 + 6 + 1 = 10$ ways to give away the 6 pencils making sure each friend has at least 1 pencil.

Problem 5.18 The Little Twelve Basketball Conference has two divisions, with six teams in each division. Each team plays each of the other teams in their own division twice and every team in the other division once. How many conference games are scheduled?

Answer

96

Solution

Let's first take a look at how many games one particular team will play. Let's call this team, "Team A". Team A will play against each of the other 5 teams in its division twice, so team A will play 10 games against teams in its division. Team A will also play once against each of the 6 teams in the other division, so that adds an extra 6 games to the 10 we already counted for a total of 16. Now, this happens for every single team in the tournament. There are 12 teams in total, so there should be $16 \times 12 = 192$ games, however, we are counting every game twice! It is the same game when we have "Team A vs. Team B" than when we have "Team B vs. Team A", so we must divide this number we got by 2. Therefore, there are $192 \div 2 = 96$ games played in total.

Problem 5.19 Henry's Hamburger Heaven offers its hamburgers with the following condiments: tomato, lettuce, pickles, and cheese. A customer can choose one, or two meat patties, and any collection of condiments. How many different hamburgers can be ordered?

Answer

32

Solution

Note there are 4 different condiments. Since a customer can choose any combination of condiments, we can think of each condiment as "it is on the hamburger" or "it is not on the hamburger". This means that for each of the 4 condiments we have 2 choices, either have it or not have it. So, there are $2 \times 2 \times 2 \times 2 = 16$ ways to choose the condiments (including the option "no condiments at all"). Since we also need to choose to have either 1 or 2 meat patties, we have in total 16 ways to choose the condiments and 2 ways to choose the meat, so $16 \times 2 = 32$ ways of ordering a hamburger.

Problem 5.20 If you have 3 nickels, 5 pennies and 1 dime. How many distinct ways can you give someone 20¢?

Answer

3

Solution

If we use all 5 nickels, we have already 15¢, so we need 5¢ more, which we can only get by using all of the pennies we have (using a dime would be too much). If we use only 2 nickels, we would have 10¢ so far, so we need 10¢ more, which we can only get by using the 1 dime as using all of our pennies would not be enough. If we use only 1 nickel, we would need to use all the pennies and the dime so cover the 15¢ missing. If we do not use any nickels there is no way we can add up to 20¢ using only the pennies and the dime. So, there are only 3 ways to give someone 20¢ using these coins.

Problem 5.21 **How many 3-digit numbers can be formed using 3, 5, and 0? Digits may not be repeated.**

Answer

4

Solution

The first digit of our number must not be 0 (otherwise it would be a 2-digit number!). We have 2 non zero digits to choose from (3 or 5). After we choose the first digit, we just need to arrange the other two digits we have left, which can be done in 2 ways. So, in total we have $2 \times 2 = 4$ different 3-digit numbers.

Problem 5.22 **How many 4-digit numbers can be formed using the digits 0, 1, 2, 3, 4 and 5? Digits may not be repeated.**

Answer

300

Solution

Our first digit must not be 0, so we have 5 ways of choosing it. Our second digit can be 0, but since we have already used one digit in the first place, we only have 5 digits left to choose from. For each of the following places in our number we have 4 and 3 possible digits to choose from, respectively, since we cannot repeat digits. This way we will have in total $5 \times 5 \times 4 \times 3 = 300$ different 4-digit numbers with no repeated digits using those digits.

Problem 5.23 **Four friends, Ivory, Katie, Nick and Peter, go to the mysterious**

pyramid to find treasures. They tried the secret code *Open Sesame* and the gate did not open, but instructions were displayed: "The secret code has to be formed using only and all of your first name initials one time". They tried a lot of different codes and they got the door open using the last code they could try, how many different codes did they check?

Answer

24

Solution

For the first letter of the code they have 4 choices, for the second letter they have 3 choices, for the third letter they have 2 choices and for the last letter they only have 1 letter left. Thus they could use $4 \times 3 \times 2 \times 1 = 24$ different codes to try to open the door.

Problem 5.24 Ursula is making up 5-letter words with some scrabble tiles she found under her sofa. She found the letters J, A, C, K and Y. How many different "words" can she make if she wants the third letter not to be Y?

Answer

96

Solution

Since she does not want the third letter to be a Y, let's start by choosing what to use in that place. There are 5 letters to choose from, so there are 4 options for what letter to use in the third place. Now we just need to place the 4 remaining letters in the other 4 places, this can be done in $4 \times 3 \times 2 \times 1 = 24$ ways. Thus, there are $4 \times 24 = 96$ ways to form letters using these tiles such that the third letter is not a Y.

Problem 5.25 You are setting up the passcode for your new phone. You have to choose 1 letter followed by 3 digits. How many codes can you make if you only want to use odd digits and you do not want to use vowels?

Answer

2625

Solution

There are $26 - 4 = 21$ letters that are not vowels and 5 odd digits. This means you could choose the code in $21 \times 5 \times 5 \times 5 = 2625$ different ways.

Problem 5.26 **At the DMV they have a new rule about license plates. Before they allowed only one letter followed by 5 digits, now they allow 2 letters followed by 4 digits. How many more license plates can they have with this new rule?**

Answer

4160000

Solution

With the old rule they could have $26 \times 10 \times 10 \times 10 \times 10 \times 10 \times 10 = 2600000$ different license plates. With the new rule they can have $26 \times 26 \times 10 \times 10 \times 10 \times 10 = 6760000$ different license plates. So, the total number of license plates increased by $6760000 - 2600000 = 4160000$.

Problem 5.27 **A grocer stacks oranges in a pyramid-like stack whose rectangular base is 5 oranges by 8 oranges. Each orange above the first level rests in a pocket formed by four oranges in the level below. The stack is completed by a single row of oranges. How many oranges are there in the stack?**

Answer

100

Solution

Each level of the pyramid of oranges will have less oranges. Every time there will be one less orange per side. The bottom layer will have 5 oranges by 8 oranges. The second layer will have 4 oranges by 7 oranges. We will continue this way until we get to the top layer of oranges which will have 1 orange by 4 oranges. This means the layers will have $5 \times 8 = 40$, $4 \times 7 = 28$, $3 \times 6 = 18$, $2 \times 5 = 10$ and $1 \times 4 = 4$ oranges, respectively, for a total of $40 + 28 + 18 + 10 + 4 = 100$ oranges.

Problem 5.28 **There are seven students and seven chairs lined up in a room. In how many ways can the students be seated in the room?**

5040

Let's start by sitting students one chair at a time. The first student to sit will have 7 chairs to choose from. The second student will have 6 chairs to choose from. The third one will have 5 chairs to chose from and so on until the last student has only one chair to choose from. This means there will be

$$7 \times 6 \times 5 \times 4 \times 3 \times 2 \times 1 = 5040$$

different ways of sitting the students in the chairs.

Problem 5.29 **In how many ways can the students from the previous problem be seated if we arrange the chairs in a circle and the oldest person has to be seated in the chair that is facing the door?**

720

This time we know that the oldest person will not have a choice and will have to sit in a particular chair. Every other person will have as many choices as chairs left in the room. This means we can sit the rest of the people in

$$1 \times 6 \times 5 \times 4 \times 3 \times 2 \times 1 = 720$$

different ways.

Problem 5.30 **When your family of 8 people gets together for Christmas, you all sit together around a rectangular table that has 8 chairs. Your grandfather always sits at the head of the table and your grandma sits to his side. In how many ways can you all choose to sit at the table?**

1440

Solution

Your grandma has 2 chairs to choose from, either to the left or to the right of your grandfather. Each of the other 6 people will be able to chose whatever chair they want that is still available when they sit. This means there are in total

$$1 \times 2 \times 6 \times 5 \times 4 \times 3 \times 2 \times 1 = 1440$$

different ways to sit for the Christmas dinner.

Problem 5.31 John is playing Monopoly with his friends. If he has four *M$3* bills and three *M$5* bills, how many different amounts of money could he pay to someone by using one or more bills?

Answer

19

Solution

$5 \times 4 - 1 = 19$

Problem 5.32 Today's Chef's Specials were served on 23 plates of different colors: 8 red plates, 6 green plates, 5 white plates, and 4 blue plates. How many of today's Chef's Specials were not served on blue plates?

Answer

19

Solution

We only have 4 blue plates, which means there were $23 - 4 = 19$ Chef's specials that were not served on a blue plate.

Problem 5.33 Pat will select three cookies from a tray containing only chocolate chip, oatmeal, and peanut butter cookies. There are at least three of each of these three kinds of cookies on the tray. How many different assortments of three cookies can be selected?

Answer

10

Solution

Pat could choose one of each cookie, two of one flavor and one of a different flavor, or all of the cookies the same flavor. She can choose one of each flavor in 1 way, two of one flavor and one of a different flavor in $3 \times 2 = 6$ ways, and all of them the same flavor in 3 ways (just choose the flavor!). Thus, in total Pat could choose the cookies in $1 + 6 + 3 = 10$ different ways.

Problem 5.34 **Your mom wants you to go and buy some lottery tickets for her. She told you to buy "all the tickets that end in** 6 **and start with** 894 **with no repeated digits". If each lottery ticket has a** 7**-digit number and they never use the digit** 1**, how many tickets can you buy?**

Answer

60

Solution

All the tickets your mom wants you to buy look like

$$894\_\_\_6$$

while not having repeated digits or using the digit 1. For the three numbers missing the digits that you still have available are 0, 2, 3, 5, and 7, that is, 5 different digits. Since you do not want to have repeated digits, there are 5 choices for the first missing digit, 4 choices for the second missing digit and 3 choices for the third missing digit. That is, in total there are $5 \times 4 \times 3 = 60$ possible tickets that you could buy.

Problem 5.35 **Daniel is about to buy a new car. He already knows the model of the car he wants to buy but he can also choose the color of the exterior and the color of the seats, as well as extra features. There are** 7 **exterior colors available,** 2 **colors for the seats, and he can choose to add GPS navigation system, satellite radio and heated seats. In how many ways can Daniel configure the car he wants to buy? (Note that he may add more than one extra feature or may not want to add any extra features at all).**

Answer

112

Solution

Let's first deal with the extra features that he can choose. Each of the features is optional, so he could either have it on the car or not have it. So, for each of the 3 extra feature he has 2 choices, that is, he has $2 \times 2 \times 2 = 8$ ways of deciding which extra features to add (including the option of not choosing any of them at all). Daniel also has to choose one of the 7 exterior colors and one of the 2 colors for the seats, so he could configure his car in $7 \times 2 \times 8 = 112$ different ways.

6. Figurate Numbers

Problem 6.1 **Identify a pattern and figure out the next number in each sequence.**

(a) $1, 4, 9, 16, 25, \ldots$

Answer

36

Solution

The numbers in the sequence are the square numbers: $1^2 = 1$, $2^2 = 4$, $3^2 = 9$, $4^2 = 16$, $5^2 = 25$, so the next number must be $6^2 = 36$.

(b) $1, 5, 14, 30, 55, \ldots$

Answer

91

Solution

Let's take a look at the difference of consecutive numbers in the sequence: $5 - 1 = 4$, $14 - 5 = 9$, $30 - 14 = 16$, $55 - 30 = 25$. Each of those differences is a square number! This means the next number must be 55 plus the next square number, so $55 + 36 = 91$.

(c) $2, 6, 12, 20, 30, \ldots$

Answer

42

Solution

Let's take a look again at the difference of consecutive numbers in the sequence: $6 - 2 = 4$, $12 - 6 = 6$, $20 - 12 = 8$, $30 - 20 = 10$. We can see that each number in the sequence is obtained by adding the next even number to the last number on the sequence: $2 + 4 = 6$, $6 + 6 = 12$. $12 + 8 = 20$, $20 + 10 = 30$, so the next number must be $30 + 12 = 42$.

Problem 6.2 **Consider the number sequence** $6, 1, 3, 6, 1, 3, 6, \ldots$ **What is the** 20th **number in the sequence?**

Answer

1

Solution

This time we can see that the numbers in the sequence repeat every 3 numbers. The remainder when 20 divided by 3 is 2, so the number in the 20th position must be the second in the repeating pattern, which is 1.

Problem 6.3 **What is the units digit of** 2^{2017}**?**

Answer

2

Solution

Let's first find a pattern in the last digits of the powers of 2:

---

power	number	last digit
2^1	2	2
2^2	4	4
2^3	8	8
2^4	16	6
2^5	32	2
2^6	64	4
2^7	128	8
2^8	256	6

We can see then that the last digits of powers of 2 follow a repeating pattern that repeats every 4 numbers: 2, 4, 8, 6, 2, ... Since when 2017 is divided by 4 the remainder is 1, so the last digit of 2^{2017} must be the first number in out repeating pattern, which is 2.

Problem 6.4 What is the next figure in the sequence?

(a) $\triangle, \square, \triangle, \square, \ldots$

Answer

$\triangle$

Solution

We have a sequence that alternates $\triangle$ and $\square$, so the next figure in the pattern must be $\triangle$.

(b) $\star\triangleleft, \star\triangleright, \star\triangleleft, \ldots$

Answer

$\star\triangleright$

Solution

So far very figure in the sequence starts with $\star$ and is followed by either $\triangleleft$ or $\triangleright$. We can see also that $\triangleleft$ and $\triangleright$ alternate in the sequence, so the next figure in the sequence should be $\star\triangleright$.

(c) $\square, \triangle\square\square, \triangle\triangle\square\square\square, \ldots$

Answer

Solution

Notice that every figure has one more △ and one more □ than the previous figure. The last figure has 2 △s and 3 □s, so the next figure must be △△△□□□□.

Problem 6.5 **I'm wearing a hat that I got** 17 **days before Tuesday. What day was that?**

Answer

Saturday

Solution

We know that the days of the week repeat every 7 days. The day 17 days before Tuesday will be the same day as $17 - 7 = 10$ days before Tuesday and also $10 - 7 = 3$ days before Tuesday. 3 days before Tuesday is Saturday, so that must be the day I got my hat.

Problem 6.6 **Identify a pattern in each sequence of triangles and draw the next triangle in the sequence.**

(a) △, △, △, △, ...

Answer

Solution

Notice that all the triangles are facing up and the bold sides follow a pattern that repeats every 3 triangles, so the next triangle must be also facing up with a bold side on the left:

(b) △, ▽, △, ▽, ...

Answer

Solution

This time our triangles are alternating between facing up and down. Compare this sequence of triangles with the one from the previous problem. The triangles that are facing up are the same as the triangles in the other sequence, and the triangles facing down are the same as in the previous sequence but rotated half a turn. The next triangle in the sequence must be facing up, so it should be the same as in the previous problem:

Problem 6.7 Lindy is about to start an internship on September 1st. If the internship lasts for 20 months, what is the last month she will be working there?

Answer

April

Solution

September is the 9th month of the year, which is 9 months from January. We want to find out what month is it if we count $20 + 9 = 29$ months starting from January. Since months repeat every 12 months, that is the same as $29 - 12 = 17$ months after January, which is also the same as $17 - 12 = 5$ months after January, which is May. Note however that we have found that 20 months after September 1st is May 1st, so the last month that Lindy will be working there will be April.

Problem 6.8 I got my new bike six days after Tuesday. What day of the week did I get my bike?

Answer

Monday

Solution

Note that the day of the week that occurs 6 days after Tuesday is the same as the day of the week that occurs 1 day before Tuesday (since the days of the week repeat every 7 days), so the day we are looking for is Monday.

Problem 6.9 Suppose yesterday was Monday. If we start from today and count the days, what day would it be on the 18th day?

Answer

Friday

Solution

Since yesterday was Monday, then today is Tuesday. 18 days counting from Tuesday is the same as $18 - 7 = 11$ days from Tuesday, which is also the same as $11 - 7 = 4$ days from Tuesday. 4 days starting to count from Tuesday will be then Friday.

Problem 6.10 The inspector came to Roy's restaurant on Monday. He said that he would be coming back 8 business days later. What day of the week is he coming back? (Business days are Monday through Friday).

Answer

Thursday

Solution

We want to be careful when counting the business days of the week. We can pretend we are on a planet that has weeks of 5 days only (Monday through Friday). 8 days after Monday will be the same as $8 - 5 = 3$ days after Monday, so the inspector will be back on Thursday.

Problem 6.11 What is the 10th number in the following sequences?

(a) $1, 2, 2, 3, 3, 3, 4, \ldots$

Answer

4

Solution

We can see that on this sequence each number is repeated as many times as the number itself, so we have one 1 followed by two 2s, three 3s, four 4s and so on. This means that the next three numbers in the sequence must all be 4 since we can only see one 4 so far. Additionally, the 10th number in the sequence is just the fourth 4, thus the answer is

4.

(b) 22, 33, 333, 44, 444, 4444, 55, ...

Answer

55555

Solution

All of the numbers in this sequence have one repeated digit. We have one number with the digit 2, two numbers with the digit 3, three numbers with the digit 4, so we must have then four numbers with the digit 5. We can also see that the first time that we use a new digit we start with a 2-digit number, then a 2-digit number until we have as many digits as the digit we are using. This means the next number (the 8th number) on the sequence must use the digit 5 three times, which is 555, and the 9th number must be 5555, and finally the 10th number is 55555

(c) 0, 1, 10, 11, 100, 101, 110, 111, ...

Answer

1001

Solution

This sequence is only using the digits 0 and 1. Note how the digit in the last place alternates from 0 to 1 every number. The next digit (from right to left) changes every 2 numbers. The third digit (from left to right) changes every 4 numbers. So the fourth digit (from right to left) should change every 8 numbers. So far we have 8 numbers on the list so the 9th number should start with a 1 followed by 0, 0 and 0, which is 1000. For the 10th number, the last place changes from 0 to 1 again, and we get 1001.

Note: These numbers are called binary numbers.

Problem 6.12 **Answer the first few questions and try to spot a pattern to answer the last question. Note: three or more points are noncollinear if you cannot draw a *straight line* that goes through more than two of them at the same time.**

(a) **How many straight lines can be drawn between 2 points?**

Answer

1

Solution

Between two points we can only draw one straight line.

(b) How many lines can be drawn between 3 **noncollinear points?**

Answer

3

Solution

If we have 3 points, we can draw a line from each of them to the other two points. Then we can draw 3 lines in total.

(c) How many lines can be drawn between 4 **noncollinear points?**

Answer

6

Solution

We have seen already that between 3 noncollinear points we can draw 3 lines. If we have 4 points we can pretend that we already had 3 and then added one more. This means we have all 3 lines we had before, and we can now draw 3 more lines from this new point, each to one of the previous 3 points we already had, for a total of $3 + 3 = 6$ lines.

(d) How many lines can be drawn between 20 **noncollinear points?**

Answer

190

Solution

If we continue adding one point at a time, we can find a pattern of the number of lines that we can draw:

Number of points	New lines	Total number of lines
2	1	1
3	2	3
4	3	6
5	4	10
6	5	15
⋮	⋮	⋮

Note that every time we get the next triangular number! So, when we have 20 points we should have the 19th triangular number, that is $19 \times 20 \div 2 = 19 \times 10 = 190$.

Problem 6.13 **Clive is stacking boxes at work. He is super meticulous and tries to follow a pattern every time he adds more boxes to the existing stack of boxes he has. He started with no boxes. His stack of boxes looked like this the first four times he added boxes to it:**

(a) **What is the pattern that Clive is following?**

Answer

Every time he makes a bigger square

(b) **How many boxes will he have in his stack after the 6th time he adds boxes to it?**

Answer

36

Solution

The pattern we have identified so far suggests that every time there will be a square when he adds more boxes to the stack. The current stack has $4 \times 4 = 16$ boxes, so the 5th stack will have $5 \times 5 = 25$ boxes and the 6th stack will have $6 \times 6 = 36$ boxes.

(c) How many boxes does he need to add to the 6th stack to get the 7th stack?

Answer

13

Solution

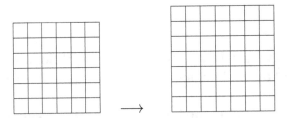

The 7th stack of boxes will have $7 \times 7 = 49$ boxes, so he will need to add $49 - 36 = 13$ boxes to the stack.

Problem 6.14 Look again at the previous problem and fill in the following table. Then answer the following questions.

Stack number	# of boxes added	total # of boxes
1	1	1
2	3	4
3	5	
4		
5		
6		
7		

(a) What pattern do you notice in the number of boxes that are being added each time the stack grows?

Answer

Add next odd number

Solution

Every time he is adding as many boxes as the next odd number.

(b) What pattern do you notice in the total number of boxes in each stack?

Answer

Square numbers

Solution

The numbers in the total number of boxes have to be the square numbers since the stacks are always squares.

(c) What is the sum of the first 100 odd counting numbers?

Answer

10000

Solution

By looking at this pattern we can see that every time Clive adds the next odd number to the stack of boxes and every time he gets in total the next square number. So, after adding boxes to the stack 100 times he would have added the first 100 odd numbers and he would get in total the 100th square number, that is $100 \times 100 = 10000$.

(d) Find the sum $11 + 13 + 15 + \cdots + 97 + 99$

Answer

2475

Solution

The sum is adding all odd numbers from 11 to 99. That is, it is adding the first 50 odd numbers except for the first 5 odd numbers. We know now that the sum of the first 50 odd numbers is $50 \times 50 = 2500$ and the sum of the first 5 odd numbers is $5 \times 5 = 25$, so the sum will be the same as $2500 - 25 = 2475$.

Problem 6.15 What is the next figure in the sequence?

(a) ♡◇, ♡◇♡◇, ♡◇♡◇♡◇, ♡◇♡◇♡◇♡◇, ...

Answer

♡◇♡◇♡◇♡◇

Solution

Every figure has 2 more symbols as the previous figure, so the next figure must have 8 symbols. Also we can see that the symbols in each figure alternate from ♡ to ◇, so the next figure should be

$$♡◇♡◇♡◇♡◇$$

(b) ⋆★⋆, ⋆★★⋆, ⋆★★★⋆, ⋆★★★★⋆, ...

Answer

⋆★★★★★⋆

Solution

Every time we have two ⋆s with some ★s in between. The number of ★s increases by one every time, so the next figure you'd have 5 ★s:

$$⋆★★★★★⋆$$

(c) ←, →, ⇇, ⇉, ←⇇, ⇉→, ⇇⇇, ⇉⇉, ...

Answer

←⇇⇇

Solution

We can see every time that we have arrows pointing left or right. The direction of the arrows alternates every time. And every time we have one more arrow than the last time the arrows were pointing in that direction. Thus, the next figure will have 5 arrows pointing to the left:

$$←⇇⇇$$

Problem 6.16 **There is a number sequence**

$$3, 9, 7, 4, 7, 2, 5, 5, 3, 9, 7, 4, 7, 2, 5, 5, 3, 9, 7, 4, \ldots$$

What is the 2017th term in the sequence?

Answer

3

Solution

If we look close at the numbers, we can see the list repeats itself every 8th number. To figure out what number will be in the 2017th position, we just need to find out the remainder of 2017 divided by 8. Since $2017 \div 8 = 252$ with a remainder of 1, the number we are looking for will be the 1st within the list that keeps repeating, which is 3.

Problem 6.17 **Fill out the missing term in the figure sequence.**

(a) ◁ ▷ ⋆∪,   ▷⋆∪◁,   _____,   ∪◁▷⋆,   ◁▷⋆∪,   ...

Answer

⋆∪ ◁▷

Solution

Note that every figure in the sequence has 4 symbols. Also the symbols appear to follow the same order, but each time we start with the second we used second in the previous figure. So, the next figure should start with ⋆, that is

⋆∪ ◁▷

(b) ↑,   ↓↑↓,   _____,   ↓↑↓↑↓↑,   ↑↓↑↓↑↓↑,   ...

Answer

↑↓↑↓

Solution

We have a pattern of alternating up and down arrows in each figure. Every time that we add arrows we add 2 on the ends facing the opposite direction as the last arrows of the previous figure. Thus the figure we are missing is

$$\uparrow\downarrow\uparrow\downarrow\uparrow$$

(c) $\star\star$, $\star\,||\,\star$, $\star\,|\,\star\star\,|\,\star$,   _____, $\star\,|\,\star\,|\,\star\star\,|\,\star\,|\,\star$,   ...

Answer

$\star\,|\,\star\,||\,\star\,|\,\star$

Solution

We have a pattern with $\star$s and $|$s. We can see that every time the next figure has 2 more symbols than before and the symbols are added in between the place where we had repeated symbols in the previous figure. The second figure has 2 $\star$s together, so the next figure should have 2$|$s together in between those, that is:

$$\star\,|\,\star\,||\,\star\,|\,\star$$

Problem 6.18 Kori was back at her mom's office. She had so much fun making patterns with sticky notes last time that she went on with it again. This time her first patterns looked like this:

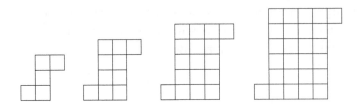

(a) **How many sticky notes would she need for her 6th figure?**

Answer

50

Solution

The 5th and 6th figures that Kori will make look like this:

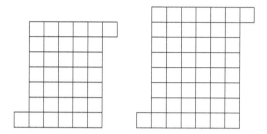

So she needs 50 sticky notes for the 6th one.

(b) Write a description that could be used to determine the shape and the total number of sticky notes in figure 6. Your description should be clear enough so that another person could read it and use it to think about another figure.

Solution

Each figure starts with a square and at the top and bottom has an extra row that is one sticky note longer than the big square.

(c) How many sticky notes would she need for the 30th figure?

Answer

962

Solution

According to the pattern we described she will start by making a square of 30 by 30 sticky notes and then put on top and bottom rectangles of 31 by 1 squares, so she will need

$$30 \times 30 + 31 + 31 = 962$$

sticky notes.

Problem 6.19 Find the missing numbers in the following pattern:

1	2	4	1
2	?	7	0
6	7	14	1
?	0	3	0

Answer

5 and 3

Solution

On the last row/column we have the number in the third row/column minus the sum of the first two rows/columns.

1	2	4	1
2	**5**	7	0
6	7	14	1
3	0	3	0

Problem 6.20 **Find the value of** $99 - 98 + 97 - 96 + \cdots + 3 - 2 + 1$

Answer

50

Solution

Notice that we can group the numbers like

$$(99 - 98) + (97 - 96) + (95 - 94) + \cdots + (3 - 2) + 1$$

so that we end up with

$$\underbrace{1 + 1 + 1 + \cdots + 1 + 1}_{50 \text{ times}} = 50$$

Problem 6.21 **What is the next figure in the sequence?**

(a) △□, △△□, △△△□□, △△△△△□□, △△△△△△△□□□, ...

Answer

△△△△△△△△△□□□□

Solution

Note that every figure starts with several △s followed by □s. Each figure has as many triangles as the total number of symbols in the previous figure, and the number of squares doubles every two figures. So, the next figure should have 11 △s followed by 4 □s:

△△△△△△△△△△△□□□□

(b) ↓, ↑↑↓↑↑, ↓↓↓↑↑↓↑↑↓↓↓, ↑↑↑↑↓↓↓↓↑↑↓↑↑↓↓↓↓↑↑↑↑, …

Answer

↓↓↓↓↓↑↑↑↑↓↓↓↑↑↑↓↑↑↑↓↓↓↑↑↑↑↓↓↓↓↓

Solution

We have an alternating pattern of upwards and downwards arrows. Every time we add arrows facing in the opposite direction as the previous time. We add arrows on both ends and we add one more arrow than in the previous figure. So, the next figure should be

↓↓↓↓↓↑↑↑↑↓↓↓↑↑↓↑↑↓↓↓↓↑↑↑↑↓↓↓↓↓

(c) ◁▷, ◁◁▷◁▷▷, ◁◁◁◁▷◁▷▷◁◁◁▷◁▷▷▷, …

Answer

◁◁◁◁▷◁▷▷◁◁▷◁▷▷▷◁◁◁◁▷◁▷▷◁◁▷◁▷▷▷▷

Solution

Note how in every figure we have two copies of the previous figure with and extra ◁ and ▷ in the end. So, the next figure is

Problem 6.22 Chandler came up with a rule to write a sequence of numbers: if the last number he wrote is even, he divides it by 2 and that is the next number in the sequence; if the last number he wrote is odd, he adds 1 to it and then divides it by 2 to come up with the next number. He decided that the first number would be 250. What is the 15th number in his sequence of numbers?

Answer

1

Solution

Following Chandler's rule, the numbers will be: $250, 250 \div 2 = 125, (125+1) \div 2 = 63,$ $(63+1) \div 2 = 32, 32 \div 2 = 16, 16 \div 2 = 8, 8 \div 2 = 4, 4 \div 2 = 2, 2 \div 2 = 1, (1+1) \div 2 = 1,$ $(1+2) \div 2 = 1, \ldots$ Note now from the 9th number we always get the number 1, so the 15th number should be a 1.

Problem 6.23 Consider the following number sequence:

$$101, 102, 103, 100, 105, 98, 107, \ldots$$

(a) What is the 10th term in the sequence?

Answer

94

Solution

If we look only at the numbers in the odd positions, we can see that they are all odd numbers and they increase by 2 every time. If we look only at the number in the even positions we can see they are all even numbers and they decrease every time by 2. The 10th number will be in an even position and should be $2 \times 4 = 8$ less than 102, so it is 94.

(b) What is the 15th term in the sequence?

Answer

115

Solution

The 15th number will be in an odd position and should be $2 \times 7 = 14$ more than 101, so it is 115.

(c) Try to find more than one way to describe the pattern the numbers are follow-

ing.

1. The odd numbered terms form an arithmetic sequence that starts with 101 and increases by 2 each time; the even numbered terms form an arithmetic sequence that starts with 104 and decreases by 2 each time.
2. Each term (except the first three) is the same as the sum of the numbers three positions before and two positions before minus the number before it ($100 = 101 + 102 - 103$, $105 = 102 + 103 - 100$, ...).

Problem 6.24 **Tom was bored while waiting for his mom to come to pick him up from school. While he waited he repeatedly wrote "TOMISAWESOMESAUCE" in a small piece of paper (leaving no spaces in between each time he wrote the phrase). He had just enough space to write 230 letters in the piece of paper. What was the last letter he wrote?**

Answer

S

Solution

The phrase Tom is writing has 17 letters. When we divide 230 by 17 we get a quotient of 13 and remainder of 9. This means Tom was able to write the phrase 13 times and then the first 9 letters of the phrase. The letter in the 9th position is S, so that is the last letter he wrote.

Problem 6.25 **Neal arrived late at his friend's birthday party and everyone was playing a game. He didn't hear the rules so he was trying to figure out what the game was about. It was weird, they were counting starting from one, but sometimes they would clap. When someone didn't clap when they were supposed to the game was over and they started again: "one", "two", *clap*. "four", "five", *clap*, "seven", "eight", *clap*, "ten", "eleven", *clap*, *clap*, "fourteen", *clap*,**
...

(a) What do you think are the rules to the game?

Answer

They must clap instead of saying the number if the number is a multiple of 3 or if the number has the digit 3 in it.

Solution

Compare the list of counting numbers with the pattern they are following:

Number	Game
1	one
2	two
3	*clap*
4	four
5	five
6	*clap*
7	seven
8	eight
9	*clap*
10	ten
11	eleven
12	*clap*
13	*clap*
14	fourteen
15	*clap*
⋮	⋮

We can see that they are clapping instead of saying numbers that are multiples of 3, but they did not clap either when it was the turn of number 13, so they must not be clapping when the number has a 3 in one of its digits.

(b) After listening his friends play a couple of rounds Neal thought he finally got how the game was played, so he joined. Right before his turn someone said "twenty". What should Neil say/do in his turn?

Answer

clap

Solution

21 is a multiple of 3, so Neil should clap.

(c) At some point during the game Neil hears them clapping ten times in a row. What was the next number they said out loud?

Answer

"forty"

Solution

Clapping 10 times in a row means they were going through the numbers 30, 31, 32, ..., 39, so the next number is 40 for which Neil should say "forty".

Problem 6.26 **Find the missing numbers in the following pattern:**

1	2	3	0
4	5	6	**?**
7	8	9	6
−2	−1	**?**	−3

Answer

3 and 0

Solution

The last row/column has the number we get by adding the numbers in the first two rows/columns and subtracting the number in the third row/column.

1	2	3	0
4	5	6	**3**
7	8	9	6
−2	−1	**0**	−3

Problem 6.27 **Find the missing numbers in the following pattern:**

1	3	4	12
2	5	?	70
6	7	2	84
12	?	56	70560

Answer

7 and 105

Solution

The last row/column has the product of the numbers in the first three rows/columns.

1	3	4	12
2	5	**7**	70
6	7	2	84
12	**105**	56	70560

Problem 6.28 Nick and two of his friends went to an amusement park and are planning on getting on a roller coaster. There are a lot of people in line and the roller coaster has 8 rows of 4 seats each, labeled A, B, C and D, that are filled in order. The first four people in line go on the first row, the next four people go on the second row, and so on. In each row, the people fill in the seats in the same order: A, B, C and D.

(a) If there are 43 people in front of Nick and his friends are right after him in line, are they going to be sitting on the same row in the rollercoaster?

Answer

No

Solution

The rollercoaster can fit $8 \times 4 = 32$ people. So a full rollercoaster will be filled with people before Nick and his friends leaving $43 - 32 = 11$ people in line before them. Then 8 people will fill rows 1 and 2, and Nick will use the last seat (D) in row 3. His friends will seat in 4A and 4B, so they will not be seating in the same row together.

(b) If two people cut in line ahead of Nick. What row and seat is Nick going to be sitting at?

Answer

4B

Solution

The two people that cut in line ahead of Nick will push him back two places, so he will now be in 4B, instead of 3D, and his friends will be next to him on 4C and 4D.

Problem 6.29 **Kori is at it again. This time she found some triangle stickers and decided to make some figures with them following a pattern. The third and fourth figures she made look like this:**

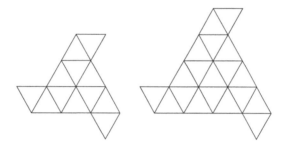

(a) What were her first two figures?

Answer

(b) Describe the pattern that Kori is following. How many triangle stickers does she use for each figure? Make sure to describe the pattern so that another person can read it and use it to think about the figure.

Answer

If she is making figure n, she uses $n \times n + 3$ triangle stickers

Solution

If we say that each triangle sticker is an equilateral triangle of side 1, she first makes a triangle with side length equal to the number of the figure she is making (so the first figure she starts with a triangle of side 1, the second time with a triangle of side 2, and so on) Note that she need as many stickers as the square of the side length for this part. Then she adds one extra sticker at the end of the bottom side, at the top of the right side and at the bottom of the left side of the big triangle. So, if she is making figure n, she uses $n \times n + 3$ triangle stickers in total.

(c) How many triangle stickers did she use for her seventh figure?

Answer

52

Solution

According to the pattern we found, she will need $7 \times 7 + 3 = 52$ stickers. We can verify our answer by looking at the actual figure:

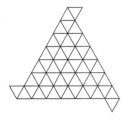

Problem 6.30 Label the balls in the pyramid with the numbers of Pascal's triangle. Note that you may only have to work hard for half of them and the rest of them will be easy to find. Can you see why?

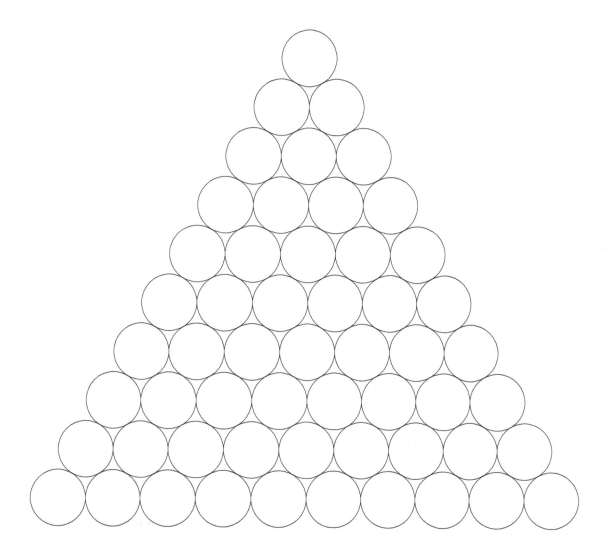

Remember we need to start by writing 1s on the top and on the beginning and end of each row, then to find any other number in the triangle we need to add up the two numbers on top of it.

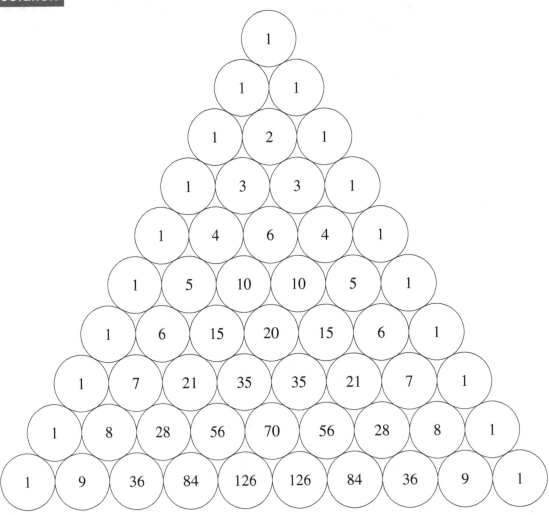

Problem 6.31 In the diagram of Pascal's triangle you filled out above, find

(a) The counting numbers.

Answer

The counting numbers are in the second diagonal (the one right next to the diagonal that has only 1s).

(b) The sequence of triangular numbers.

Answer

All of the triangular numbers can be found in the third diagonal of the triangle, right below the counting numbers.

Problem 6.32 **Look again at the diagram of Pascal's triangle.**

(a) Find two balls that are touching each other that add up to $2^2 = 4$.

Answer

1 and 3 in the third and fourth layers

(b) Find two balls that are touching each other that add up to $3^2 = 9$.

Answer

3 and 6 in the fourth and fifth layers

(c) Find two balls that are touching each other that add up to $4^2 = 16$.

Answer

6 and 10 in the fifth and sixth layers

(d) Use the balls you identified to figure out a pattern to find the square numbers in Pascal's triangle.

Answer

The square numbers can be found by adding consecutive triangular numbers (the ones that we found were on the third diagonal of the triangle).

Problem 6.33 **We will use once more the diagram you figured out for Pascal's triangle. For each of the following questions we will add some of the numbers in the triangle. It might be great if you color the number we use in each question with a different color so you can identify a pattern. Start by coloring the top ball with one color and the first ball of the second layer with a different color.**

(a) Add the numbers in the first ball of the third layer and the second ball of the second layer (use another color to paint these two balls).

Answer

2

Solution

$1+1=2$

(b) Add the numbers in the first ball of the fourth layer, and the second ball of the third layer (use another color to paint these two balls).

Answer

3

Solution

$1+2=2$

(c) Add the numbers in the first ball of the fifth layer, the second ball of the fourth layer and the third ball of the third layer (use another color to paint these three balls).

Answer

5

Solution

$1+3+1=5$

(d) Add the numbers in the first ball of the sixth layer, the second ball of the fifth layer and the third ball of the fourth layer (use another color to paint this three balls).

Answer

8

Solution

$1+4+3=8$

(e) Following this pattern, what numbers should you add up next? What is their sum?

Answer

$1 + 5 + 6 + 1 = 13$

Solution

We need to add the numbers in the first ball of the seventh layer, the second ball in the sixth layer, the third ball in the fifth layer and the fourth ball in the fourth layer. Note that every time we start one layer below and we move one ball to the right when we go up in each layer until we can't move up anymore.

(f) Continue following the pattern and make a list of the numbers you add up in each step. Keep that list and come back to this problem after you go through the problems in the next chapter. You'll see then why we wanted to follow this pattern.

Answer

1, 1, 3, 5, 8, 13, 21, 34, 55, ...

Solution

These are the Fibonacci numbers! (If you have not seen these number yet, you will in the next chapter!)

Problem 6.34 In the following diagrams label the triangles that are pointing upwards with the numbers of Pascal's triangle, starting with a 1 on the uppermost triangle. Use a color to fill in the triangles that have odd numbers (odd numbers are those that aren't divisible by 2). See if you can identify a pattern before you color the last diagram.

(a)

(b)

(c)

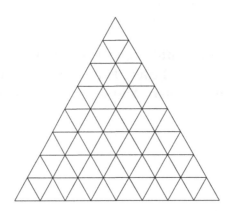

(d)

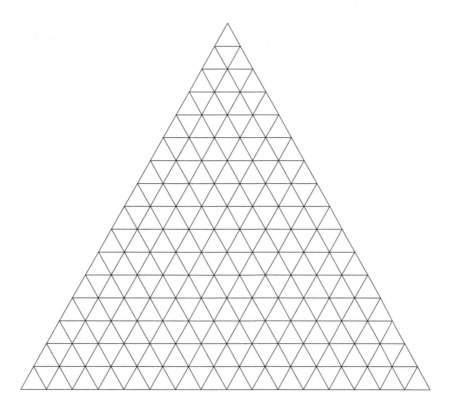

Answer

Problem 6.35 In the following diagrams label the triangles that are pointing upwards with the numbers of Pascal's triangle, starting with a 1 on the uppermost triangle. Use a color to fill in the triangles that have numbers that are not divisible by 3. See if you can identify a pattern before you color the last diagram.

(a)

(b)

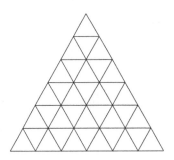

(c)

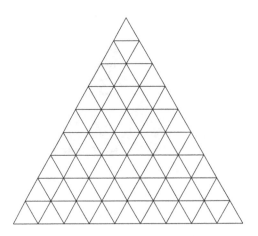

(d)

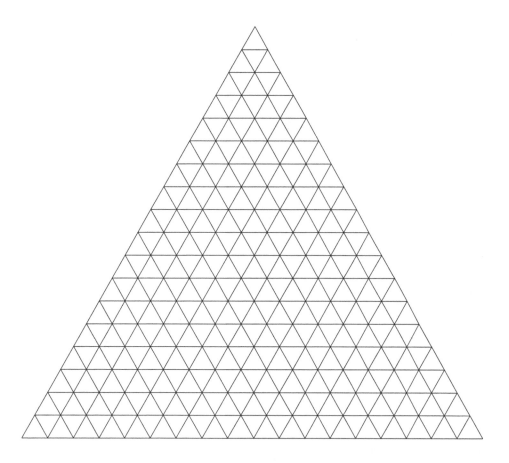

Answer

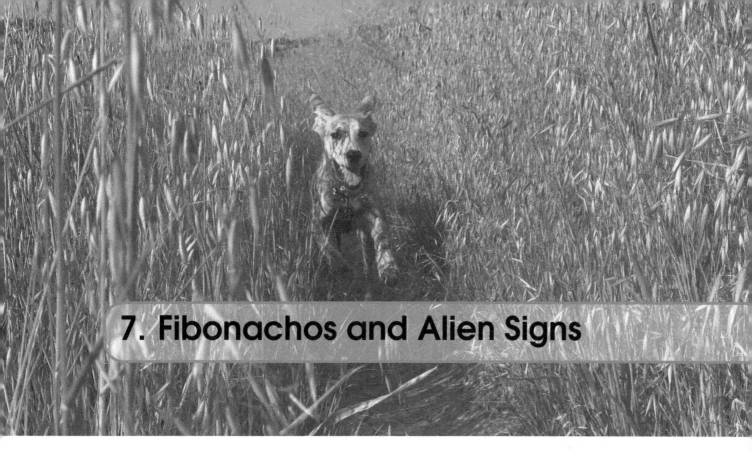

7. Fibonachos and Alien Signs

Problem 7.1 **Find the sum of all the counting numbers from 1 to 10.**

Answer

55

Solution

The counting numbers from 1 to 10 form an arithmetic sequence that increases by 1 and has 10 terms, so their sum will be

$$\frac{(1+10) \times 10}{2} = 55$$

Problem 7.2 **If we start with 3 and count every fifth number, what will be the tenth number we count?**

Answer

48

Solution

We have an arithmetic sequence that increases by 5 starting with 3. To get to the tenth number in the list we need to add 5 nine times, so we would get to the number

$$3 + 5 \times 9 = 48.$$

We can see that the first 10 numbers in our list would be

$$3 \quad 8 \quad 13 \quad 18 \quad 23 \quad 28 \quad 33 \quad 38 \quad 43 \quad 48$$

Problem 7.3 **If we start with** 3 **and we keep multiplying by** 11**, what would be the fourth number in our list?**

Answer

3993

Solution

This time we have a geometric sequence that starts with 3 and has ratio 11. The first few terms of our sequence will be

$$3 \quad 3 \times 11 = 33 \quad 33 \times 11 = 363 \quad 363 \times 11 = 3993$$

so the fourth number in our list would be 3993.

Problem 7.4 **What is the next number in the sequence?**

(a) $5, 8, 11, 14, \ldots$

Answer

17

Solution

If we subtract consecutive terms in our sequence we can see that we have a common difference of 3:

$$8 - 5 = 3 \quad 11 - 8 = 3 \quad 14 - 11 = 3$$

So the next number is $14 + 3 = 17$.

(b) 21, 26, 31, 36, ...

Answer

41

Solution

This time we have a common difference of 5:

$$26 - 21 = 5 \quad 31 - 26 = 5 \quad 31 - 26 = 5 \quad 36 - 31 = 5$$

So the next number is $36 + 5 = 41$.

(c) 125, 150, 175, 200, ...

Answer

225

Solution

The common difference this time is 25:

$$150 - 125 = 25 \quad 175 - 150 = 25 \quad 200 - 175 = 25$$

The next number on the list is $200 + 25 = 225$.

Problem 7.5 **What is the sum of the first 10 even counting numbers?**

Answer

110

Solution

The first 10 even counting numbers form an arithmetic sequence that increases by 2. The 1st even counting number is 2 and the 10th even counting number is 20, so the sum of the first 10 even counting numbers is

$$\frac{(2 + 20) \times 10}{2} = 22 \times 5 = 110$$

Problem 7.6 What is the next term in the sequence?

(a) $1, 10, 100, 1000, 10000, \ldots$

Answer

100000

Solution

This time we have a geometric sequence as we get the same quotient every time we divide two consecutive terms in the sequence:

$$10 \div 1 = 10 \quad 100 \div 10 = 10 \quad 1000 \div 100 = 10 \quad 10000 \div 1000 = 10$$

The first term is 1 and the common ratio is 10, so the next term in the sequence is $10000 \times 10 = 100000$.

(b) $1, 4, 16, 64, 256, \ldots$

Answer

1024

Solution

We have again a geometric sequence, this time the first term is again 1 and the common ratio is 4:

$$4 \div 1 = 4 \quad 16 \div 4 = 4 \quad 64 \div 16 = 4 \quad 256 \div 64 = 4$$

So the next term in the sequence is $256 \times 4 = 1024$.

(c) $1, 11, 121, 1331, 14641, \ldots$

Answer

161051

Solution

We can see this time the common ratio of the geometric sequence is 11:

$$11 \div 1 = 11 \quad 121 \div 11 = 11 \quad 1331 \div 121 = 11 \quad 14641 \div 1331 = 11$$

So the next term in the sequence is $14641 \times 11 = 161051$.

Problem 7.7 Find the next number in each of the following sequences:

(a) $3, 7, 10, 17, 27, \ldots$

Answer

44

Solution

Let's take a look at the difference of consecutive terms in the sequence:

$$7 - 3 = 4 \quad 10 - 7 = 3 \quad 17 - 10 = 7 \quad 27 - 17 = 10$$

Can you see a pattern? Note how the second, third and fourth differences correspond to the first, second and third terms of the original sequence. This tells us that this is a sequence like the Fibonacci sequence, where each term is the sum of the previous two, but it starts with 3 and 7 instead of 1 and 1. This way, the terms of the sequence are:

$$3 \quad 7 \quad 3+7 = 10 \quad 7+10 = 17 \quad 10+17 = 27 \quad 17+27 = 44,$$

so the next number in the sequence is 44.

(b) $5, 2, 7, 9, 16, \ldots$

Answer

25

Solution

We have one more time a "Fibonacci-like" sequence, this time we start with 5 and 2 and each of the next terms is the sum of the previous two:

$$5 \quad 2 \quad 5+2 = 7 \quad 2+7 = 9 \quad 7+9 = 16 \quad 9+16 = 25,$$

so the next number in the sequence is 25.

(c) $2, 3, 5, 8, 13, \ldots$

Answer

21

Solution

This is a "Fibonacci-like sequence" that starts with 2 and 3, so the first few terms are:

$$2 \quad 3 \quad 2+3=5 \quad 3+5=8 \quad 5+8=13 \quad 8+13=21 \quad 13+21=34$$

The next term in the sequence is 21. Note this is the same as the original Fibonacci Sequence skipping the first two terms.

Problem 7.8 **The average of three consecutive counting numbers is 55. What is the smallest of the three numbers?**

Answer

54

Solution

Consecutive counting numbers form an arithmetic sequence that increases by 1. Whenever we have an odd number of terms in an arithmetic sequence the term in the middle is the average of all of the numbers. We know the average of 3 numbers is 55, so the middle number must be 55. This means the three numbers in our arithmetic sequence are

$$54, \quad 55, \quad 56.$$

So the smallest of them is 54.

Problem 7.9 **What is** $13 \times 13 - 12 \times 12$ **equal to?**

Answer

25

Solution

We know that the sum of the first 13 odd numbers $(1 + 3 + \cdots + 25)$ is the same as 13×13, and the sum of the first 12 odd numbers $(1 + 3 + \cdots + 23)$ is the same as 12×12, so this difference should be equal to just the 13th odd number, so 25.

Problem 7.10 **What is the sum of the first 8 Fibonacci numbers?**

Answer

54

Solution

The sum of the first 8 Fibonacci numbers will be the same as 1 less than the 10th Fibonacci number. The first 10 Fibonacci numbers are

$$1, \quad 1, \quad 2, \quad 3, \quad 5, \quad 8, \quad 13, \quad 21, \quad 34, \quad 55$$

So the sum of the first 8 Fibonacci numbers is $55 - 1 = 54$.

Problem 7.11 Sammy Sandstone was taking some pictures on a field trip to the savanna. At first he could only spot one lion, so he took his first picture. Two more lions approached and he took a second picture, capturing all three lions at once. It seems that more lions were arriving, two at a time, and Sammy took a picture every time there were new lions on the field.

(a) How many lions did Sammy have in his seventh picture?

Answer

13

Solution

The number of lions in Sammy's pictures follow an arithmetic sequence that starts with 1 and increases by 2 each time. The 7th picture will have $6 \times 2 = 12$ more lions than the first picture, so $1 + 12 = 13$ lions in total.

(b) How many lions did Sammy have in all seven of his pictures?

Answer

49

Solution

Since the number of lions follow an arithmetic sequence, and we know that the first picture had 1 and the 7th had 13 lions, then between all 7 pictures there will be

$$\frac{(1 + 13) \times 7}{2} = 49$$

lions.

Problem 7.12 **Jen, my dog, used to like eating my socks. From January to June she ate 54 of my socks. I know that each month she ate 2 more than the previous month. How many socks did she eat in June?**

Answer

14

Solution

Since Jen eats 2 more socks each month, we know that the total number of socks she ate in January plus June is the same as February plus May and is also the same as March plus April. So, Jen must have eaten a total of

$$54 \div 3 = 18$$

socks in March and April combined. Since we also know that Jen ate 2 more socks in April, she must have eaten 8 socks in March and 10 in April. So she ate $10 + 2 = 12$ socks in May and $12 + 2 = 14$ socks in June.

Problem 7.13 **3 letters arrived in the mail on Monday. 5 letters arrived in the mail on Tuesday, and 7 arrived on Wednesday. If this pattern continues until Sunday. How many pieces of mail will have accumulated by Sunday?**

Answer

63

Solution

It seems the mail is arriving according to an arithmetic sequence that starts with 3 and increases by 2 each time. Sunday comes 6 days after Monday, so on Sunday $3 + 6 \times 2 = 15$ letters will arrive. Since we have 7 days from Monday to Sunday, the total number of letters that arrived is

$$\frac{(3 + 15) \times 7}{2} = 63.$$

Problem 7.14 **Corry went to the forest every day for a week to collect rocks. The first day he collected 5 rocks. Each of the following days he collected 2 more rocks than the previous day. What is the average of the number of rocks he collected each day?**

Answer

11

Solution

The number of rocks that Corry collected form an arithmetic sequence. We know that the average of an odd number of terms in an arithmetic sequence is the same as the middle number in the sequence. So, in average he will collect as many rocks as he collected on the 4th day of the week (the 4th is the middle day among 7 days). On the 4th day he collected $3 \times 2 = 6$ more rocks than the first day, that is $5 + 6 = 11$ rocks. Thus, Corry collected 11 rocks in average each day.

Problem 7.15 **Five successive counting numbers have a sum of** 70, **what are the numbers?**

Answer

12, 13, 14, 15, 16

Solution

The average of 5 numbers is given by their sum divided by 5. We know the sum of these 5 numbers add to 70, so their average is $70 \div 5 = 14$. We also know that these are consecutive numbers, so they form an arithmetic sequence that increases by 1 each time. The average of an odd number of terms in an arithmetic sequence is the middle number of the sequence, so 14 is the number in the middle of the sequence of 5 numbers. This means the numbers are

$$12, \quad 13, \quad 14, \quad 15, \quad 16.$$

Problem 7.16 **The aliens did it again! This time Old Mr. Farmer Guy came out to his field and found a big number** 8 **in the middle of it surrounded by a circle.**

(a) **Fill in the table: Into how many parts is the number** 8 **separated if Mr. Farmer**

Guy passes 2, 3, 4 **and** 5 **with his tractor truck vertically close to the middle of the number** 8?

# of passes	1	2	3	4	5
# of parts	2				

Answer

5, 8, 11, 14

Solution

Notice how after the 1st pass the number 8 is split into 2 parts, and after that every pass increases the number of parts by 3:

# of passes	1	2	3	4	5
# of parts	2	5	8	11	14

so the number of parts in which the number 8 is divided forms an arithmetic sequence that starts with 2 and increases by 3 each time.

(b) Into how many parts is the 8 **separated if Mr. Farmer Guy passes with his tractor truck** 40 **times?**

Answer

119

Solution

We know now that the number of parts the 8 is separated into form an arithmetic sequence that starts with 2 and increases by 3 every time. So, after 40 passes there will be $39 \times 3 = 117$ more parts than the 1st time, so $2 + 117 = 119$ parts in total.

(c) If the number 8 **is now divided into** 152 **parts, how many times did Mr. Farmer Guy pass with his tractor truck?**

Answer

51

Solution

We know that 152 must be 2 plus a bunch of 3s added together,

$$152 = 2 + \underbrace{3 + 3 + \cdots + 3}_{? \text{ times}}$$

as we started with 2 parts and every time we increase the number of parts by 3.

Notice that

$$150 = 2 + 50 \times 3,$$

so we need to add 50 times the number 3 to the number 2 to get 152. This means the tractor must have passed over the field $50 + 1 = 51$ times.

Problem 7.17 Suppose you have a secret. On Sunday you tell it to a friend. On Monday your friend tells your secret to 2 other friends. On Tuesday, each of your friends who heard your secret on Monday tell it to 2 other friends. If this procedure continues, how many people in all will have been told your secret by the end of the following Sunday? Assume no one is told your secret more than once.

Answer

255

Solution

Note that every day there will be 2 times as many people knowing the secret as the day before. So, the number of new people that hear your secret each day follow a geometric sequence that starts in 1 (Sunday) and is multiplied by 2 each time. Take a look at the following table to see how many new people hear your secret each day.

Day	# of new people
Sunday	1
Monday	2
Tuesday	4
Wednesday	8
Thursday	16
Friday	32
Saturday	64
Sunday	128
Total	255

So by the end of the following Sunday a total of 255 different people know your secret.

Problem 7.18 Henry used his two favorite numbers to write a sequence like the Fibonacci sequence in a piece of paper (that is, he added the two numbers to get a third number, then he added the second number and this new third number to get the fourth, and so on). His best friend, Eustace, has always been curious about Henry's favorite numbers so he took the piece of paper with the list of numbers to find out what they are, only to discover that Henry's dog had eaten the first few numbers off the list. Eustace could see that the 5th number on the list was 109 and the 6th number on the list was 167. Can you help him figure out Henry's favorite numbers?

Answer

44 and 7

Solution

We can use the numbers to trace back the previous numbers on the list. Since the 6th number is the same as the 4th number plus the 5th number and so on. So, the 4th number on the list is $167 - 109 = 58$; the 3rd number is $109 - 58 = 51$; the 2nd number is $58 - 51 = 7$; and the 1st number is $57 - 7 = 44$. So, Henry's favorite numbers are 44 and 7.

Problem 7.19 Randia was visiting a friend at his house but he was late getting back from school. While Randia waited for her friend she decided to add up the numbers of the 7 houses in her friend's street. The house numbers are all consecutive and she got a sum of 2205. What are the numbers of the houses?

Answer

312, 313, 314, 315, 316, 317 and 318

Solution

All 7 numbers are consecutive, so they form an arithmetic sequence. the average of the 7 numbers will be the middle number of the sequence and is also equal to $2205 \div 7 = 315$. So, the numbers of the houses are

$$315 - 3 = 312, \quad 315 - 2 = 313, \quad 315 - 1 = 314, \quad 315,$$

$$315 + 1 = 316, \quad 315 + 2 = 317, \quad 315 + 3 = 318$$

Problem 7.20 **The sum of** 6 **consecutive even numbers is** 234, **what is the smallest of the numbers?**

Answer

34

Solution

Consecutive even numbers form an arithmetic sequence that increases by 2 every time. The average of the 6 numbers is $234 \div 6 = 39$. Since we found the average of an even number of terms in an arithmetic sequence, this means 39 is right in between the two middle terms of the sequence. Since the sequence increases by 2 every time, 39 is 1 more than the 3rd term and 1 less than the 4th term of the sequence, so the 3rd and 4th numbers in the sequence are $39 - 1 = 38$ and $39 + 1 = 40$. So, the rest of the numbers on the sequence of 6 numbers are

$$34, \quad 36, \quad 38, \quad 40, \quad 42, \quad 44,$$

and the smallest of the numbers is 34.

Problem 7.21 **A list contains** 7 **consecutive even numbers. If their sum is** 294, **what is the second number on the list?**

Answer

38

Solution

Once more we know the sum of the 7 numbers. We can use this sum to find their average, which is $294 \div 7 = 42$. Since we have 7 numbers, the average is the number right in the middle of the list, which is the 4th number on the list. This means the 2nd number is $2 + 2 = 4$ less than 42, so $42 - 4 = 38$.

Problem 7.22 **The following is an arithmetic sequence with some numbers missing. What are the missing numbers?**

$$14, \underline{\qquad}, \underline{\qquad}, 26, \ldots$$

Answer

18 and 22

Solution

It takes 3 steps to get from 14 to 26 on the arithmetic sequence. This means we need to add 3 times the common difference of the sequence to the number 14 to get to the number 26. The difference between 26 and 14 is $26 - 14 = 12$, which should be 3 times the common difference. This means the common difference in the arithmetic sequence is $12 \, div \, 3 = 4$, so the missing numbers are $14 + 4 = 18$ and $18 + 4 = 22$.

Problem 7.23 **The fourth and sixth terms of a geometric sequence are** 135 **and** 1215. **Suppose the terms of this sequence are all positive, what is the fifth term of the sequence?**

Answer

405

Solution

It takes 2 steps to get from the fourth to the sixth term of a geometric sequence. This means we need to multiply 135 by the same number twice to get to 1215. The quotient between 1215 and 135 is $1215 \div 135 = 9$, which should be the common ratio multiplied by itself. Since the terms are all positive, this means the common ratio in the geometric sequence is 3. So, the sixth number in the sequence is $135 \times 3 = 405$.

Problem 7.24 **Find the sum of all the counting between the indicated limits:**

(a) 3 **through** 12

Answer

75

Solution

Remember that consecutive numbers form an arithmetic sequence that increases by 1. To find these sums we just need to add the first and last numbers on the sequence, multiply by the number of terms in the sequence, and divide by 2. From 3 to 12 we have

10 numbers, so the sum of the numbers from 3 to 12 is equal to

$$(3+12) \times 10 \div 2 = 75.$$

(b) 15 **through** 55

Answer

1435

Solution

From 15 to 55 we have $55 - 15 + 1 = 41$ numbers, so the sum of the numbers from 15 to 55 is

$$(15+55) \times 41 \div 2 = 1435.$$

(c) 5 **through** 90

Answer

4085

Solution

From 5 to 90 we have $90 - 5 + 1 = 86$ numbers, so the sum is equal to

$$(5+90) \times 86 \div 2 = 4085.$$

Problem 7.25 **The third term in a geometric sequence is** 20, **and the fourth term is** 40. **What is the sixth term?**

Answer

160

Solution

The numbers 20 and 40 are consecutive in the geometric sequence. So, we can find the common ratio of the sequence if we divide $40 \div 20 = 2$. So, the fifth term is $40 \times 2 = 80$ and the sixth term is $80 \times 2 = 160$.

Problem 7.26 **The following is a geometric sequence with some numbers missing. What are the missing numbers?**

$$12, \underline{\hspace{1cm}}, \underline{\hspace{1cm}}, 324, \ldots$$

Answer

36 and 108

Solution

It takes 3 steps to get from 12 to 324 in the sequence, that means $324 \div 12 = 27$ is equal to the common ratio multiplied by itself three times. Since $3 \times 3 \times 3 = 27$, the common ratio of the sequence is 3. This means the second number is $12 \times 3 = 36$ and the third number is $36 \times 3 = 108$.

Problem 7.27 **Ashley wants to find the average temperature in a week. She checked the thermometer outside her window the first thing every morning, and recorded the following table:**

Monday	Tuesday	Wednesday	Thursday	Friday	Saturday	Sunday
66°	67°	70°	69°	71°	68°	72°

What is the average temperature in the week?

Answer

69°

Solution

To add the numbers Ashley decided to first order them from smallest to largest:

$$66, 67, 68, 69, 70, 71, 72$$

and she noticed that they were consecutive numbers. Ashley knows that the average of an odd number of consecutive numbers is the number in the middle of the list, so the average must be 69 If ordered, the numbers form an arithmetic sequence. We have 7 numbers and the number in the middle is 69.

Problem 7.28 Jing and Xing went to the Fibonachos restaurant and ordered a plate of 54 nachos for them to eat together. Jing ate one nacho and then they took turns eating.

(a) How many nachos did Jing eat?

Answer

21

Solution

Remember, in the restaurant they must each nachos following the Fibonacci sequence. The plate had 54 nachos and since the first 8 Fibonacci numbers add up to

$$1+1+2+3+5+8+13+21 = 54,$$

each of them had 4 bites of nachos. Jing ate the first nacho, so she ate as many nachos as the odd terms in the Fibonacci sequence, that is, she ate $1+2+5+13 = 21$ nachos in total.

(b) How many nachos did Xing eat?

Answer

33

Solution

Xing ate as many nachos as the even terms in the Fibonacci sequence, so he ate $1+3+8+21 = 33$ nachos.

(c) They wondered if they could have eaten the nachos in a different order (so not alternating) and still each of them eat the same number of nachos as they did before. Is this possible? (Remember: since they ordered a single plate, each time either of them eats nachos, they have to eat as many as the next Fibonacci number on the list).

Answer

Yes

Solution

If Xing eats $1+1+2+3+5+8+13 = 33$ nachos, he would have eaten the same 33 nachos as before, and there would have been 21 nachos left for Jing to eat.

Problem 7.29

(a) What is $51 \times 51 - 50 \times 50$ **equal to?**

Answer

101

Solution

Remember the sum of consecutive odd numbers starting from 1 is equal to a square number. The sum of the first 51 odd numbers is equal to 51×51 and the sum of the first 50 odd numbers is equal to 50×50, so $51 \times 51 - 50 \times 50$ is equal to the 51st odd number, so $51 \times 2 - 1 = 101$.

(b) What is $49 \times 49 - 48 \times 48$ **equal to?**

Answer

97

Solution

This time we will have the difference between "the sum of the first 49 odd numbers" and "the first 48 odd numbers", so exactly the 49th odd number, which is $49 \times 2 - 1 = 97$.

(c) Calculate $51 \times 51 - 50 \times 50 + 49 \times 49 - 48 \times 48 + \cdots + 11 \times 11 - 10 \times 10$

Answer

1281

Solution

If we group the numbers, we can see that

$$(51 \times 51 - 50 \times 50) + (49 \times 49 - 48 \times 48) + \cdots + (11 \times 11 - 10 \times 10),$$

is the same as the sum

$$101 + 97 + \cdots + 21$$

which is the sum of numbers in an arithmetic sequence that starts in 21, increases by 4 and has 21 numbers. So, this sum is equal to

$$(101 + 21) \times 21 \div 2 = 1281.$$

Problem 7.30 Kathleen loves visiting her grandma because she has one of those grandfather's clocks that has a pendulum and chimes at the start of every hour. The clock chimes as many times as the hour. If Kathleen was at her grandma's house from 4:30 pm to 10:30 pm, how many chimes did she hear?

Answer

45

Solution

Kathleen heard the clock chime at 5:00, 6:00, 7:00, 8:00, 9:00 and 10:00. So, she heard the clock chime a total of $5 + 6 + 7 + 8 + 9 + 10 = 45$ times.

Problem 7.31 Darren was counting by threes starting at 5. His brother Phillip was adding up all the numbers. If If Darren counted 12 numbers, what sum did Phillip get?

Answer

258

Solution

Since Darren started counting at 5, the 12th number he counted was $5 + 11 \times 3 = 38$. The sum of this numbers is $(5 + 38) \times 12 \div 2 = 43 \times 6 = 258$.

Problem 7.32 The sum of ten consecutive even numbers is 370. What is the biggest number that was added?

Answer

46

Solution

When we have a list of 10 consecutive even numbers, the first and tenth number will add up to the same as the second and ninth, third and eighth, fourth and seventh, and fifth and sixth. So 370 is equal to 5 times the same number. $370 \div 5 = 74$, so each pair of numbers must add up to 74. The fifth and sixth numbers are only 2 apart, so they must be 36 and 38. This means the numbers in the list were:

$$28, 30, 32, 34, 36, 38, 40, 42, 44, 46$$

so the biggest number that was added was 46.

Problem 7.33 Manny was practicing to be a magician. He got an egg at the magician's store but he dropped it through a flight of stairs. The magic egg didn't break, but it split into three eggs and every time the eggs reached the next step each of them would split into three eggs. How many eggs could Manny see at each of the 4 steps the egg fell through?

Answer

$3, 9, 27$ and 81

Solution

On the first step the egg split in 3, so Manny would have 3 eggs in total. The next step each of the 3 eggs would split in 3, so Manny would have $3 \times 3 = 9$ eggs total. The third step each of the 9 eggs would split again in 3 for a total of $9 \times 3 = 27$ eggs. On the fourth and final step, each of the 27 eggs Manny has now would once more split in 3 for a total of $27 \times 3 = 81$ eggs.

Problem 7.34 What is the next term in each of the following sequences?

(a) $20, 60, 180, 540, \ldots$

Answer

1620

Solution

If we divide consecutive terms in the sequence, we can see that there is a common ratio:

$$60 \div 20 = 3, \quad 180 \div 60 = 3 \quad 540 \div 180 = 3$$

so this is a geometric sequence with common ratio 3 that starts in 20. So, the next term of the sequence is

$$540 \times 3 = 1620.$$

(b) 15, 60, 240, 960, ...

Answer

3840

Solution

We have once more a geometric sequence, as we find a common ratio when we divide consecutive numbers:

$$60 \div 15 = 4, \quad 240 \div 60 = 4, \quad 960 \div 240 = 4$$

so the next term in the sequence is $960 \times 4 = 3840$.

(c) 34, 170, 850, 4250, ...

Answer

21250

Solution

This time, when we divide consecutive terms in the sequence we get

$$170 \div 34 = 5, \quad 850 \div 170 = 5, \quad 4250 \div 850 = 5$$

so we have again a common ratio, which means this is a geometric sequence. The next number in the sequence is then $4250 \times 5 = 21250$.

Problem 7.35 Fill in the blank with the missing term in the sequence.

(a) 4, 12, 36, _____, 324, ...

Answer

108

Solution

We can see that this is a geometric sequence with common ratio $12 \div 4 = 3$ that starts in 4. The missing term is then $36 \times 3 = 108$.

(b) _____, 308, 2156, 15092, ...

Answer

44

Solution

If we divide consecutive terms in this sequence we can see that we have a geometric sequence:

$$2156 \div 308 = 7, \quad 15092 \div 2156 = 7.$$

To get from the first to the second term of the sequence we would then multiply by 7. To go backwards we can divide by 7. So, the first term of the sequence is $2156 \div 7 = 44$.

(c) 12, 132, _____, 15972, 175692, ...

Answer

1452

Solution

The numbers in the sequence are quite large (and growing fast!) this suggests that this could be a geometric sequence. If we divide consecutive terms in the sequence we get

$$132 \div 12 = 11, \qquad 175692 \div 15972 = 11$$

so we have a geometric sequence that starts with 12 and has common ratio 11. The missing number in the sequence is then $132 \times 11 = 1452$.

8. Chickens and Rabbits

Problem 8.1 Michelle counts her chickens and rabbits, and there are 16 heads and 46 feet. How many of each type are there?

Chickens: 9; Rabbits: 7

Solution

There are 16 animals in total. Imagine first all the animals are chickens. Then there would be

$$16 \times 2 = 32$$

total feet, which is

$$46 - 32 = 14$$

less than the actual number of feet. Since each rabbit has 2 more feet than a chicken, we need to replace

$$14 \div 2 = 7$$

chickens with rabbits. Therefore, Michelle has 7 rabbits and

$$16 - 7 = 9$$

chickens.

Problem 8.2 **In a farm there are goats and ducks. The total number of heads is 100, and the total number of legs is 296. How many animals of each type are there?**

Answer

Goats: 48; Ducks: 52

Solution

Using our imagination, pretend all the ducks stand on one foot and all the goats stand up on their back 2 legs. There are still 100 heads so 100 animals, but now only half of the legs,

$$296 \div 2 = 148,$$

of the legs are on the ground. Each animal has one leg on the ground, and each goat has one extra leg on the ground. Since each duck now has one leg on the ground, the extra

$$148 - 100 = 48$$

legs come from goats. Thus there are 48 goats and

$$100 - 48 = 52$$

ducks on the farm.

Problem 8.3 **Seventy vehicles (cars and motorcycles) are parked in a parking lot. Totally there are 190 wheels. Given that a car has 4 wheels and a motorcycle has 2 wheels, how many cars and motorcycles each are in the parking lot?**

Answer

Cars: 25; Motorcycles: 45

Solution

If all seventy vehicles where motorcycles, we would have a total of

$$70 \times 2 = 140$$

wheels in the parking lot. We know there are actually 190 wheels in the lot,

$$190 - 140 = 50$$

more wheels than if we only have motorcycles. Switching a motorcycle for a car gives 2 extra wheels, therefore we must replace

$$50 \div 2 = 25$$

motorcycles with cars to get the correct number of wheels in the parking lot. Hence there are 25 cars and

$$70 - 25 = 45$$

motorcycles in the parking lot.

Problem 8.4 **Teachers and students from the Areteem Summer Camp visited the museum. They bought a total of 100 tickets for 220 dollars. If each teacher ticket costs 4 dollars, and each student ticket costs 2 dollars, how many teachers and students were there respectively?**

Answer

Teachers 10; Students: 90

Solution

If instead we have 100 unsupervised students with no teachers, the 100 tickets would cost a total of

$$100 \times 2 = 200$$

dollars. This is

$$220 - 200 = 20$$

dollars cheaper than the actual 100 tickets. If we remove one student ticket and add one teacher ticket, it costs $2 extra. Since $20 extra was paid over the price of 100 student tickets, there must be

$$20 \div 2 = 10$$

teachers. Hence there are

$$100 - 10 = 90$$

students.

Problem 8.5 **There are some chickens and some rabbits on a farm. Suppose there are 45 heads and 136 feet in total among the chickens and rabbits, how many of the animals are chickens? How many are rabbits?**

Answer

Chickens: 22; Rabbits: 23

Solution

We use a creative way to look at this problem. Chickens only have 2 feet each, but they also have 2 wings. If we pretend the wings were also "feet", then each chicken would have 4 feet, and there would be a total of

$$45 \times 4 = 180$$

feet. However, there are in fact only 136 feet, and so

$$180 - 136 = 44$$

are extra. What are these extra "feet"? The wings! So there are 44 wings, and since each chicken has 2 wings, the number of chickens is

$$44 \div 2 = 22.$$

As a result, the number of rabbits is

$$45 - 22 = 23.$$

Problem 8.6 **Two trucks dump dirt of 500 cubic meters. Truck A carries 8 cubic meters per load. Truck B carries 4 cubic meters per load. The dirt is removed after 80 loads. How many loads are carried by truck A?**

Answer

45

Solution

If we first assume all 80 loads are all from Truck B, there would be a total of

$$80 \times 4 = 320$$

cubic meters of dirt removed. In actuality,

$$500 - 320 = 180$$

more cubic meters of dirt were removed. This means that Truck A must have taken some of the loads of dirt. Each load from Truck A has

$$8 - 4 = 4$$

more cubic meters than that from Truck B, so if we take

$$180 \div 4 = 45$$

loads from truck B and give them to Truck A, we will have the correct amount of dirt. Hence truck A carries 45 loads of dirt.

Problem 8.7 **Each set of chess is played by** 2 **students, and each set of Chinese Checkers is played by** 6 **students. A total of** 30 **sets of chess and Chinese Checkers are played by exactly** 140 **students in a school event. How many sets of each game are there?**

Answer

Chess: 10; Chinese Checkers: 20

Solution

Two students can play with one set of chess, so if all 30 sets are chess sets, then only

$$30 \times 2 = 60$$

students can participate in the school event. Every set we switch from chess to Chinese Checkers allows for

$$6 - 2 = 4$$

more students to participate. We know 140 students played in the event, which is

$$140 - 60 = 80$$

more than can play with only chess sets. Therefore we need to swap

$$80 \div 4 = 20$$

of the chess sets for Chinese Checkers sets. Hence there are

$$30 - 20 = 10$$

chess sets and 20 Chinese Checkers sets at the event.

Problem 8.8 Use 400 matches to make triangles and pentagons. Totally 90 triangles and pentagons are made with no matches left over. How many of each shape are made?

Answer

Pentagons: 65; Triangles: 25

Solution

Every triangle has 3 sides, so needs 3 matches. To make 90 triangles we then need

$$90 \times 3 = 270$$

matches. A pentagon has 5 sides, so using

$$5 - 3 = 2$$

extra matches we can turn one of the triangles into a pentagon. There are

$$400 - 270 = 130$$

extra matches, so we can turn

$$130 \div 2 = 65$$

triangles into pentagons. Thus there are

$$90 - 65 = 25$$

triangles and 65 pentagons.

Problem 8.9 Candace scored 40 points in her school's playoff basketball game. She made a combination of 2-point shots and 3-point shots during the game. If she made a total of 15 shots, how many 3-point shots did she make?

Answer

10

Solution

If all of Candace's shots were 2-point shots, she would have scored

$$15 \times 2 = 30$$

points. She actually scored

$$40 - 30 = 10$$

more points. Each 3-point shot is worth an extra point over a 2-point shot, so 10 of Candace's shots must have been 3-point shots.

Problem 8.10 There are 48 **tables in a restaurant. Small tables can seat** 2 **people, and big tables can seat 5 people. They can accommodate a maximum number of** 156 **people. How many small tables and how many big tables are there?**

Answer

Small tables: 28; Big tables: 20

Solution

If all 48 tables are small tables, the restaurant could fit a maximum of

$$48 \times 2 = 96$$

people. This is

$$156 - 96 = 60$$

fewer people than the restaurants actual capacity. Each big table can accommodate

$$5 - 2 = 3$$

more people than the small table. Now

$$30 \div 3 = 20,$$

so if we switch 20 of the small tables to big tables the restaurant can accommodate 156 people as needed. Hence there are

$$48 - 20 = 28$$

small tables and 20 big tables at the restaurant.

Problem 8.11 The counselor brought his 67 **students to the lake to go rowing,** 6 **people for each big boat and** 4 **people for each small boat. They rented** 14 **boats to fit everyone with no empty seats. How many big and small boats each did they rent?**

Answer

Big boats: 6; Small boats: 8

Solution

Remember that the counselor is a person too! This means there are a total of 68 people who fit in the boats. With 14 small boats, there is only room for

$$14 \times 4 = 56$$

people, so the counselor must have rented some big boats. The counselor needs room for

$$68 - 56 = 12$$

more people. Since a big boat carries

$$6 - 4 = 2$$

more people than a small boat, switching

$$12 \div 2 = 6$$

of the small boats to big boats will make sure the 14 boats fit everyone. This means that the counselor rented 6 big boats and

$$14 - 6 = 8$$

small boats.

Problem 8.12 **There are 25 questions in a math competition. 10 points are given to each correct answer, and 4 points are taken for each incorrect answer or unanswered question. Jeff received 180 points in the competition. How many questions did he answer correctly?**

Answer

20

Solution

Since each question is worth 10 points, if Jeff got every question correct he would have received a score of

$$25 \times 10 = 250.$$

For every question Jeff gets wrong, he loses the 10 points for a correct answer and an additional 4 points for the incorrect answer, a total decrease of

$$10 + 4 = 14$$

points. Since his actual score is

$$250 - 180 = 70$$

points less than a perfect score, this means Jeff answered

$$70 \div 14 = 5$$

questions incorrectly. Hence he got

$$25 - 5 = 20$$

questions correct in the competition.

Problem 8.13 **The company Green Pilots is organizing a company picnic at the beach. They want to save energy by carpooling to the beach site. They are able to fit 535 people into 95 vehicles that are either 5-seat sedans or 7-seat minivans. How many sedans and how many minivans do they need to use to take everyone to the picnic?**

Answer

65 sedans, 30 minivans

Solution

If the company uses 95 minivans, they can transport

$$95 \times 7 = 665$$

people to the beach. This is

$$665 - 535 = 130$$

more than they need to take, so they can switch some of the minivans for more energy efficient sedans. Each sedan holds

$$7 - 5 = 2$$

fewer people than a minivan. Therefore, they can switch

$$130 \div 2 = 65$$

minivans for sedans and still take everyone to the picnic. In the end Green Pilots can use 65 sedans and

$$95 - 65 = 30$$

to take everyone to the beach.

Problem 8.14 **A large family of 30 people goes to a restaurant. They each order either pizza or salad. The pizza costs $11.00 and salad costs $6.00. In all the family spends $265.00. How many pizzas and how many salads did the family order?**

Answer

Pizza: 17; Salads: 13

Solution

Had the whole family ordered salad, they would have spent a total of $30 \times 6 = 180$ dollars, which was

$$265 - 180 = 85$$

dollars less than they actually spent. Switching from salad to pizza cost an extra

$$11 - 6 = 5$$

dollars. Therefore, if

$$85 \div 5 = 17$$

people switched their order from salad to pizza the family would spend a total of $265. Hence

$$30 - 17 = 13$$

ordered salad and 17 ordered pizza.

Problem 8.15 **Sasha takes a mathematics competition. There are a total of 15 problems. For each correct answer, competitors receive 5 points. For each wrong answer, they instead get 2 points taken away. Sasha has 33 points in total. How many problems did she answer correctly?**

Answer

9

Solution

If Sasha answered all 15 questions correct she would receive a perfect score of

$$15 \times 5 = 75.$$

She actually has only 33 points, so

$$75 - 33 = 42$$

points were deducted. Since 2 point is taken away for each wrong answer, each wrong answer actually decreases the score by

$$5 + 2 = 7$$

points. The number of wrong answer is then

$$42 \div 7 = 6,$$

and hence

$$15 - 6 = 9$$

is the number of questions Sasha answered correctly.

Problem 8.16 Aria has 20 **coins that are nickels and dimes. The total value is** $1.40. **How many nickels and dimes does Aria have? (A nickel is** 5 **cents, and a dime is** 10 **cents.)**

Answer

Nickels: 12; Dimes: 8

Solution

It will be easiest if we convert everything from dollars to cents. Aria has a total of 1.4 dollars or

$$1.4 \times 100 = 140$$

cents. Assuming all 20 coins are nickels, there would be total of

$$20 \times 5 = 100$$

cents,

$$140 - 100 = 40$$

cents less than the actual amount Aria has. Each dime is worth

$$10 - 5 = 5$$

more cents than a nickel, so trading one nickel for a dime results in an increase of 5 cents. Therefore replacing

$$40 \div 5 = 8$$

nickels with dimes, we get the correct value of 140 cents. Hence Aria has 8 dimes and the other

$$20 - 8 = 12$$

coins are nickels.

Problem 8.17 **There are 150 birds and cats. There are 120 more bird legs than cat legs. How many birds and how many cats are there?**

Answer

Birds: 120; Cats: 30

Solution

There are 120 more bird legs than cat legs. Since each bird has 2 legs, if we remove

$$120 \div 2 = 60$$

birds then we are left with

$$150 - 60 = 90$$

animals and there are an equal number of bird and cat legs. Since each cat has twice as many legs as a bird, there must be half as many birds as cats. Therefore there are

$$90 \div 3 = 30$$

cats and

$$30 \times 2 = 60$$

of the remaining animals are birds. Hence there are

$$60 + 60 = 120$$

birds in total.

Problem 8.18 A group of 85 people rent 25 go-kart cars of two kinds at a race-track. The first kind has a capacity of 3 people and costs $60 per car. The second has a capacity of 4 people and costs $75 per car. The 85 people exactly fill all vehicles. What is the total cost in renting the 25 go-kart cars?

Answer

$1650

Solution

If all 25 go-karts are the first kind with a capacity of 3 people, the go-karts will fit

$$25 \times 3 = 75$$

people, meaning that

$$85 - 75 = 10$$

people will be left out. The second kind of go-kart can hold

$$4 - 3 = 1$$

extra person, so if we switch 10 of the go-kart to the second kind everyone will fit. Hence we have

$$25 - 10 = 15$$

go-kart of the first kind and 10 of the second. Since the first kind costs $60 per car and the second kind costs $75, the total cost is

$$15 \times 60 + 10 \times 75 = 1650$$

dollars to rent the 25 go-kart cars.

Problem 8.19 There are many ducks and sheep in a farm. If we count the heads, there are a total of 60 heads. If we count the legs, there are 96 more legs from sheep than from ducks. How many ducks and how many sheep are there in the farm?

Answer

24 ducks; 36 sheep

Solution

If all 60 animals are sheep, there are

$$60 \times 4 = 240$$

sheep legs and 0 duck legs, so there are 240 more sheep legs than duck legs. If we replace one sheep with one duck we remove 4 sheep legs and add 2 duck legs, so the difference between the number of sheep legs and duck legs reduces by

$$4 + 2 = 6.$$

The difference we want is 96, which is

$$240 - 96 = 144$$

less than if we have all sheep. Thus we need to replace

$$144 \div 6 = 24$$

sheep with ducks. Hence there are

$$60 - 24 = 36$$

sheep and 24 ducks.

Problem 8.20 **The Math Club collected donations from 52 people who live in either Landover or Salisbury. Each person from Landover contributed \$10, and each person from Salisbury contributed \$7. In total, \$44 more were collected from Landover than from Salisbury. How many people contributed from each city?**

Answer

Landover: 24; Salisbury: 28

Solution

If all the people were from Landover, the Math Club would have collected \$0 from Salisbury and

$$52 \times 10 = 520$$

from Landover. This is a difference of \$520 dollars. Every person who is from Salisbury instead of Landover changes this difference by

$$10 + 7 = 17$$

because there is \$10 less from Landover and \$7 more from Salisbury. If we know that \$44 more was collected from City A, which is

$$520 - 44 = 476$$

less the amount if all the people were from Landover, we need to swap

$$476 \div 17 = 28$$

people from Landover to Salisbury. Hence there are

$$52 - 28 = 24$$

people from Landover and 28 from Salisbury.

Problem 8.21 **In a farm the total number of chickens and rabbits is** 80. **If the number of chicken feet is** 70 **more than the number of rabbit feet, how many chickens and rabbits are there respectively?**

Answer

Chickens: 65; Rabbits: 15

Solution

If there are only chickens on the farm, there are 80 chickens and therefore

$$80 \times 2 = 160$$

chicken feet. Since there are no rabbits, there are 160 more chicken feet than rabbit feet. If we replace one chicken with a rabbit, we remove 2 chicken feet and add 4 rabbit feet, a total change in

$$2 + 4 = 6$$

of the difference between chicken and rabbit feet. Since we want this difference to be 70 and

$$160 - 70 = 90,$$

we need to replace

$$90 \div 6 = 15$$

chickens with rabbits. Therefore the farm has

$$80 - 15 = 65$$

chickens and 15 rabbits.

Problem 8.22 **A turtle has** 4 **legs and a crane has** 2 **legs. There are totally** 110 **heads of turtles and cranes, and there are** 40 **more crane legs than turtle legs. How many of each animal are there?**

Answer

Turtles: 30; Cranes: 80

Solution

There are 40 more crane legs than turtle legs. Each turtle has 4 legs, so if we assume

$$40 \div 4 = 10$$

additional turtles are added, then there are

$$110 + 10 = 120$$

heads altogether and the number of crane legs and turtle legs are the same. Since each turtle has twice as many legs as a crane, there must be half as many turtles as cranes. Since

$$120 \div 3 = 40$$

there are 40 turtles and

$$2 \times 40 = 80$$

cranes now. Removing the added 10 turtles, the original numbers are

$$40 - 10 = 30$$

turtles and 80 cranes.

Problem 8.23 **On a good day, Chris the Squirrel picks 30 hazelnuts. On a rainy day he only picks 15 hazelnuts. During a few consecutive days he picked a total of 225 hazelnuts with an average of 25 per day. How many days were rainy?**

Answer

3

Solution

First note that since Chris picked 25 hazelnuts on average and we know that he picked 225 hazelnuts in total, there must have been

$$225 \div 25 = 9$$

days in total. If each of the 9 days were sunny, Chris would have picked

$$9 \times 30 = 270$$

total hazelnuts,

$$270 - 225 = 45$$

more than he actually did. On a rainy day Chris picks

$$30 - 15 = 15$$

fewer hazelnuts than on a sunny day. Therefore,

$$45 \div 15 = 3$$

of the 9 days must have been rainy.

Problem 8.24 **Morgan needed** 85 **sticks for a project at school. Each stick is either** 4 **inches long or** 7 **inches long and the total length of all the sticks combined is** 430 **inches. How many** 4 **inch sticks and** 7 **inch sticks are there?**

Answer

4 inch: 55; 7 inch: 30

Solution

If all the sticks were 4 inch sticks, the total length of the 85 sticks would be

$$85 \times 4 = 340$$

inches. Since the actual length is

$$430 - 340 = 90$$

inches longer, some of the 4 inch sticks need to be replaced with 7 inch sticks. Each 7 inch stick is

$$7 - 4 = 3$$

inches longer. Since the actual length is 90 more than the length of all 4 inch sticks, we must replace

$$90 \div 3 = 30$$

4 inch sticks with 7 inch sticks. Hence there are

$$85 - 30 = 55$$

four inch sticks and 30 seven inch sticks.

Problem 8.25 **A spider has** 8 **legs and no wings. A dragonfly has** 6 **legs and** 2 **pairs of wings. A cicada has** 6 **legs and one pair of wings. There are a total of** 27 **bugs of these types, with** 180 **legs and** 28 **pairs of wings in total. How many dragonflies are there?**

Answer

10

Solution

There are 27 bugs in total. Spiders have no wings, so let's first focus on the number of legs. If all the bugs are spiders, there are a total of

$$27 \times 8 = 216$$

legs, which is

$$216 - 180 = 36$$

more than the actual amount. Each spider has

$$8 - 6 = 2$$

more legs than a dragonfly or cicada, so there must be a combined total of

$$36 \div 2 = 18$$

dragonflies and cicadas. Each cicada has 1 pair of wings and each dragonfly has 2 pairs of wings, so since we need 28 wings in total, we must have

$$28 - 18 = 10$$

extra pairs of wings, so a total of 10 dragonflies.

Problem 8.26 **Lily spent** $465 **to buy** 67 **color pencils for her art class, including red, green, and blue colors. The red pencils cost** $3 **each, the green ones cost** $10 **each, and the blue ones cost** $7 **each. Suppose she bought the same number of green and blue pencils. How many of each type of pencils did she buy?**

Answer

Red: 19; Green: 24; Blue 24

Solution

Since Lily bought the same number of green and blue pencils, we can pretend that one green pencil and one blue pencil come bundled as a green-blue pair that costs

$$10 + 7 = 17$$

dollars. If we suppose all the pencils Lily bought were red, then the total cost would have been

$$67 \times 3 = 201$$

dollars. The actual amount 465 is

$$465 - 201 = 264$$

dollars more. Replacing a pair of red pencils, which costs

$$2 \times 3 = 6$$

dollars, with a green-blue pair costs, which costs 17 dollars, is

$$17 - 6 = 11$$

dollars more expensive. Therefore the total number of green-blue pairs is

$$264 \div 11 = 24.$$

Thus there were 24 green and 24 blue pencils. The remaining

$$67 - 24 - 24 = 19$$

are red pencils.

Problem 8.27 A crab has 10 legs. A mantis has 6 legs and 1 pair of wings. A dragonfly has 6 legs and 2 pairs of wings. There are a total of 44 of the three types. There are 312 legs in total. There are 43 pairs of wings in total. How many of each kind are there?

Answer

Crabs: 12; Mantises: 21; Dragonflies: 11

Solution

If all 44 animals are mantises, there will be a total of

$$44 \times 6 = 264$$

legs, which is

$$312 - 264 = 48$$

legs more than the actual amount. Each dragonfly also has 6 legs, but each crab has

$$10 - 6 = 4$$

extra legs. Hence there must be

$$48 \div 4 = 12$$

crabs to give us the 48 extra legs. The remaining

$$44 - 12 = 32$$

are either mantises or dragonflies. They each have the same number of legs, so we need to look at the pairs of wings. If we again assume all 32 remaining are mantises, there will be a total of

$$32 \times 1 = 32$$

pairs of wings, which is

$$43 - 32 = 11$$

less than the actual amount. Each dragonfly has

$$2 - 1 = 1$$

extra pair of wings, so there must be 11 dragonflies in total. Lastly, the remaining

$$32 - 11 = 21$$

are mantises. In all there are 12 crabs, 21 mantises, and 11 dragonflies.

Problem 8.28 **Tony's mom took out $570 from the bank. There are $2, $5, and $10 bills and total of 120 bills. The number of $5 bills and $10 bills is the same. How many bills of each type are there?**

Answer

$2: 60; $5: 30; $10: 30

Solution

We know there are equal numbers of $5 and $10 bills, so pairs these bills up for a total of

$$5 + 10 = 15$$

dollars per pair. If Tony's mom has only $5 and $10 bills, there will be

$$120 \div 2 = 60$$

pairs for a total of

$$60 \times 15 = 900$$

dollars, which is

$$900 - 570 = 330$$

more than she actually has. A pair of $2 bills is worth

$$15 - 2 \times 2 = 11$$

dollars less than a $5 and $10 bill pair, so there must be

$$330 \div 11 = 30$$

pairs of $2 bills. Hence there are

$$60 - 30 = 30$$

$5 and $10 bill pairs. Tony's mom therefore has

$$30 \times 2 = 60$$

$2 bills, 30 $5 bills, and 30 $10 bills.

Problem 8.29 **60 mice eat 60 cakes. If each big mouse eats 4 cakes, and 4 baby mice eat 1 cake, how many big mice and baby mice are there?**

Answer

Big mice: 12: Baby mice: 48

Solution

We can group them and then solve the problem. Let's combine 1 big mouse and 4 baby mice into a group, then in this group, there are 5 mice and they will eat

$$4 + 1 = 5$$

cakes. Since there are 60 cakes in total, there must be

$$60 \div 5 = 12$$

groups. In each group, there is only one big mouse, so the total number of big mice is

$$12 \times 1 = 12.$$

In each group, there are 4 baby mice, so there are

$$12 \times 4 = 48$$

baby mice in total.

Problem 8.30 A candy shop sells three flavors of candies: cherry, strawberry, and watermelon. The prices are \$15/kg, \$20/kg, and \$25/kg, respectively. This morning the shop sold a total of 105 kg and received \$2150. Given that the total sale of cherry and watermelon flavor candies combined was \$1250, how many kilograms of watermelon flavor candies were sold?

Answer

35

Solution

We know the total sale was \$2150. Since \$1250 of this was for cherry and watermelon, the remaining

$$2150 - 1250 = 900$$

dollars were due to strawberry. Hence

$$900 \div 20 = 45$$

kilograms of strawberry candies was sold. Thus the other

$$105 - 45 = 60$$

kilograms were cherry and watermelon. If these 60 kg were all cherry, the total sales would be

$$60 \times 15 = 900$$

dollars, which is

$$1250 - 900 = 350$$

less than the actual amount. As each kilogram of watermelon is

$$25 - 15 = 10$$

dollars more expensive, there must have been

$$350 \div 10 = 35$$

kg of watermelon flavor candies sold.

Problem 8.31 John and Jane are holding a party at their home in the jungle. For the party, they need to collect 200 bananas. John can collect 35 bananas per hour, while Jane can collect 20 bananas per hour. Further, only one of them can collect bananas at a time, as someone must stay at home to watch their son Korak. John starts collecting the bananas and then Jane takes over to collect the rest. If a total of 7 hours is spent collecting bananas, how many hours are spent by Jane?

Answer

3

Solution

If John was the only one collecting bananas for the entire 7 hours, he would collect

$$35 \times 7 = 245$$

bananas. That is

$$245 - 200 = 45$$

more bananas than they actually collected. Jane can collect

$$35 - 20 = 15$$

less bananas per hour than John, so she must have collected bananas for

$$45 \div 15 = 3$$

hours.

Problem 8.32 Jason and Perry work together to cut the grass at their school's soccer field. In total, the area of the field is 280 square yards. Jason can mow 60 square yards per hour and Perry can mow 40 square yards per hour. They start moving the lawn together, but Jason quits before the whole field is complete and Perry finishes the job working a total of 4 hours. How many hours did Jason work?

Answer

2

Solution

If Perry had worked alone for 4 hours he would have mown

$$4 \times 40 = 160$$

square yards. That is

$$280 - 160 = 120$$

less yards than they needed to mow. This means Jason mowed the rest, so he spent

$$120 \div 60 = 2$$

hours mowing the lawn.

Problem 8.33 **Fred and Ted are in charge of painting desks for Mrs. Larson's classroom, they want all the desks to be painted white. There are 30 desks in total, 19 of them are already white and the rest are gray. On the first day, only Ted painted, but he got confused, and started painting the already white desks gray. On the second day, only Fred painted, and by the end of the day all 30 desks (including the ones Ted had painted the wrong color) were white. Assume that Ted can paint 1 desk per hour and Fred can paint 2 desks per hour. If we know that altogether Ted and Fred painted for 16 hours, how many hours did Fred paint?**

Answer

9

Solution

Before they started painting, there were $30 - 19 = 11$ gray desks. If Fred alone had painted desks for 16 hours, he would have painted

$$2 \times 16 = 32$$

desks, that is

$$32 - 11 = 21$$

more desks than they needed to paint. As Fred paints 2 desks per hour and Ted "unpaints" 1 desk per hour, Fred can paint

$$2 + 1 = 3$$

more desks per hour than Fred. So, Ted spent

$$21 \div 3 = 7$$

hours painting desks and Fred spent

$$16 - 7 = 9$$

hours painting desks.

Problem 8.34 Maxime must drive 12 miles to work. Some of the trip is on the highway, where it takes him 1 minute to drive 1 mile. The rest of the trip is on city streets, where it takes 2 minutes to drive 1 mile. If his whole trip takes 20 minutes, how many miles does he drive on the highway? on city streets?

Answer

Highway: 4 miles; City Streets: 8 miles

Solution

If Maxime had used only the high way to drive to work, he would have spent 12 minutes to get there, that is,

$$20 - 12 = 8$$

less minutes than the time he actually spent. For every mile he drives on city streets, he spends

$$2 - 1 = 1$$

more minute on the road. So he must have driven on city streets for 8 miles and for

$$12 - 8 = 4$$

miles on the highway.

Problem 8.35 Alice completed 15 laps at the local track. Some of the laps she walked and the others she ran. Running, it takes Alice 2 minutes to complete a lap. Walking, it takes Alice 5 minutes to complete a lap. Disappointed in her time, the next day Alice completes the laps again, this time running every lap she walked the first time and walking every lap she ran the first time. This improves Alice's time by 15 minutes. How many laps did Alice run originally?

Answer

5

It takes Alicia

$$5 - 2 = 3$$

more minutes to walk a lap than to run a lap. On the second day, she swapped running and walking in every lap, improving her time by 15 minutes, so she must have walked

$$15 \div 3 = 5$$

laps this time and ran

$$15 - 5 = 10$$

laps. So, originally Alice ran for 5 laps and walked for 10 laps.